AT BUS DESIGN

Compatible with IEEE P996
8 and 16 Bit ISA, E-ISA, and EISA Design

Edward Solari

Annabooks

AT BUS DESIGN

by

Edward Solari

PUBLISHED BY

Annabooks
12145 Alta Carmel Court, Suite 250
San Diego, CA. 92128

Printed in the United States of America

ISBN 0-929392-08-6

First Printing March 1990
Second Printing July 1990
Third Printing February 1991
Fourth Printing October 1991

Dedication

To my mother and father for my education,
and to my wife for her patience.

Acknowledgements

The Author and the Publisher would like to acknowledge the co-operation of the IEEE P996 Working Group. Scott Hopkinson of Emulex and Mike Fung of Chips and Technologies and were particularly helpful in obtaining permission of the IEEE to quote portions of the P996 proposed standard.

The A.C. timing tables and signal waveform drawings in this book represent a composite of information from several sources in addition to the information that is original to this book. The following sources have provided portions of the A.C. timing and signal waveforms:

"ISA BUS SPECIFICATIONS and APPLICATION NOTES, January 30, 1990", Copyright 1989, Intel Corporation

"TECHNICAL REFERENCE GUIDE, EXTENDED INDUSTRY STANDARD ARCHITECTURE EXPANSION BUS" Copyright COMPAQ COMPUTER Corp., 1989

"PERSONAL COMPUTER BUS STANDARD P996" DRAFT D2.00, January 18, 1990, Copyright IEEE Inc., 1990.

Cover design by Amy Cooper.

AT BUS DESIGN

Contents

Page

List of Figures

List of Figures (Continued)

Page

List of Figures (Continued)

Page

List of Tables

List of Tables (Continued)

List of Tables (Continued)

Foreword

I met Ed Solari at a meeting of the IEEE P996 Working Group. The objective of the Working Group was to publish a definitive standard for the AT (ISA) Bus, and I joined the group near the end, when most of the work was done. It didn't take many meetings, however, for me to realize that the task was sizeable and complex; I became very impressed with the caliber of people who were devoting their time to this complicated task.

At the time, I was also in search of someone to write this book. I have been involved in non-desktop applications of PCs since 1982, and have watched many designs take place without the foundation of a formal standard. Now, being in the technical publishing business, it seemed natural to fill the documentation void that had existed for so long.

I met at least four people on the Working Group that might have written this book. Of the four, Ed stood out for two reasons: he is doubly qualified by virtue of his work on both the ISA and EISA Busses, and (believe it or not) he was able to take time off from work to write. And here, only a few months later, is the result.

The evolution of this bus from the original PC to the present ISA, E-ISA, and EISA varieties could easily cause so much confusion that a designer would despair of ever really having full confidence in his design. However, Ed has given this potential mess considerable thought, and has done a masterful job of organizing his work. We hope you will agree, and that you will find this book both useful and easy to use.

John P. Choisser
Annabooks

Preface

The purpose of this book is to document the AT bus and the ISA portion of the EISA bus. The information contained in this book is for engineers designing add-on cards or compatible platforms, and technicians needing to repair or trouble-shoot existing products.

The information contained in this book represents not only a consolidation of recently published standards, but also contains information derived from independent research and personal insights on how AT busses work. This information is of particular importance because IBM has never made their AT bus specification available to outsiders. Many engineers have "reversed-engineered" the AT bus, resulting in a diversity of interpretations. In addition, platform design engineers have had to replicate all of the design features and flaws of the AT platform.

Most recently the EISA consortium has released a bus specification that is an "extension" of the AT bus. This, in conjunction with two additional bus specifications from the IEEE and Intel Corporation, provides the basic information to allow the AT bus to be fully documented. In addition, EISA specifies enhancements to the basic AT (ISA) bus. In order to prevent confusion between the "traditional" and the "enhanced" AT/ISA bus, this book has adopted the E-ISA bus concept. The E-ISA bus is simply the ISA bus with the appropriate 8/16 bit enhancements of the EISA bus.

This book does not deal with the full 32 bit EISA bus. It has purposely focused on the bus designs that support AT, ISA, and E-ISA compatible 8 and 16 bit add-on cards. This focus was needed to prevent confusion with the radically different operation of the 32 bit EISA bus. Add-on cards that are AT, ISA, or E-ISA compatible will work with AT, ISA, or EISA platforms. In addition, the E-ISA compatible add-on cards can exploit additional features available on EISA compatible systems.

CHAPTER ONE

ARCHITECTURAL OVERVIEW

INTRODUCTION

There are now three basic iterations of the IBM Personal Computer: the original PC , the XT, and the AT. The platform CPU for the PC and XT is the 8 bit Intel 8088, and for the AT it is the 16 bit Intel 80286. The platforms vary in other ways besides the processor, such as memory capacity, existence of an audio cassette interface, and the number of slots for the add-on cards. There are two versions of the add-on card slots: the 8 bit slots on the PC and XT, and the 8 and 8/16 bit slots on the AT.

The PC and XT both use the Intel 8284A chip for clock generation, the Intel 8288 chip for bus control, a 4.77 Mhz platform CPU and bus clock, and the Intel 8237A-5 DMA controller. The operation of the 8 bit slots on the two platforms is identical, and establishes the standard 8 bit compatibility for processor access cycles and DMA transfers. Both platforms use one of the DMA channels to execute refresh cycles on the bus.

The AT differs from the PC and XT in other ways besides processor type and the data width of the bus slots. First, the AT uses the Intel 82284 chip for clock generation and the Intel 82288 chip for bus control. Second, the AT comes in two versions with two different system and platform CPU clocks: 6 and 8 Mhz. Third, additional DMA channels and interrupt signal lines are present on the bus. Fourth, add-on cards can replace the platform CPU as the bus owner. Fifth, an additional signal line is available to support faster no wait state cycles. Finally, the refresh cycle execution by the DMA controller on the PC and XT is replaced by discrete circuitry.

The differences in the AT relative to the PC and XT are further complicated because companies other than IBM have replaced the platform CPU with the Intel 80386 or 80486 CPUs. The use of these CPUs eliminates the use of the Intel bus controller and clock generation chips, and permits the use of other platform CPU clock frequencies.

In order to permit the development of add-on cards in some reasonable fashion, the IEEE, Intel, and a consortium of companies (lead by COMPAQ) have developed three sets of bus specifications. The IEEE and Intel have focused on the 8 and 8/16 bit slots of the AT as the standard. The COMPAQ-lead consortium (known as EISA) also focused on the 8 and 8/16 bit slots of the AT. However, EISA enhanced the bus specification beyond the AT. IBM has never released documentation describing the operation of the AT bus.

This book is focused on the AT bus specifications as documented by the IEEE and EISA. In that these specifications use the IBM/AT as a standard, they are all in excellent agreement. The generic name for the 8 bit and 8/16 bit AT compatible bus is INDUSTRY STANDARD ARCHITECTURE (ISA). The generic name for the 32 bit extension to the ISA standard is EXTENDED INDUSTRY STANDARD ARCHITECTURE (EISA).

In developing the EISA bus specification, the COMPAQ-lead consortium added additional features beyond the increase of the data width to 32 bits. Some of these features can be used by ISA platforms that only support 8 and 8/16 bit slots. Thus, for the purposes of this book, the following definitions will be used:

An ISA platform refers to the basic AT compatible (IEEE P996) 8 or 8/16 slot bus specification. The bus clock speed is a maximum of 8 Mhz. It does not use any of the enhancements of the EISA bus specification. It does not have the extra connector pins to support the 32 bit EISA bus cycles.

An E-ISA platform is the same as an ISA platform, except it includes those EISA enhancements that do not require EISA cycles or 8/16/32 bit EISA slots. The bus clock speed is a maximum of 8.33 Mhz. An E-ISA platform can be viewed as the 8 and 8/16 bit slots of an EISA(32) platform that uses only ISA compatible signal lines.

An EISA(32) platform conforms to the full EISA bus specification, which includes all of the ISA and E-ISA features. It also contains extra circuitry and connector pins to support EISA cycles and 8/16/32 bit EISA slots.

ISA PLATFORM ARCHITECTURE

The ISA platform is the basic model for all PC, XT, and AT platforms. (See Figure 1-1.) The basic platform consists of the following:

PLATFORM CPU

This is the primary CPU of the platform. It is the first bus owner after power up and after every reset. All other bus owners must obtain bus ownership by halting code execution by the platform CPU. It is directly tied to the clock circuitry which drives the bus clock; consequently it is the only resource that can use the no wait state access cycles. For software compatibility, it must be compatible to the INTEL 8088, 80286, 80386, and 80486 family of processors.

PLATFORM MATH CO-PROCESSOR

This resource is a companion to the platform CPU. It is an option that computes math-related functions faster than the platform CPU. It is invisible to the slot in terms of bus cycles.

MEMORY

Memory resources can exist either on the platform or on a bus add-on slave card. Platform memory can be of any byte width compatible to the platform CPU, provided the platform

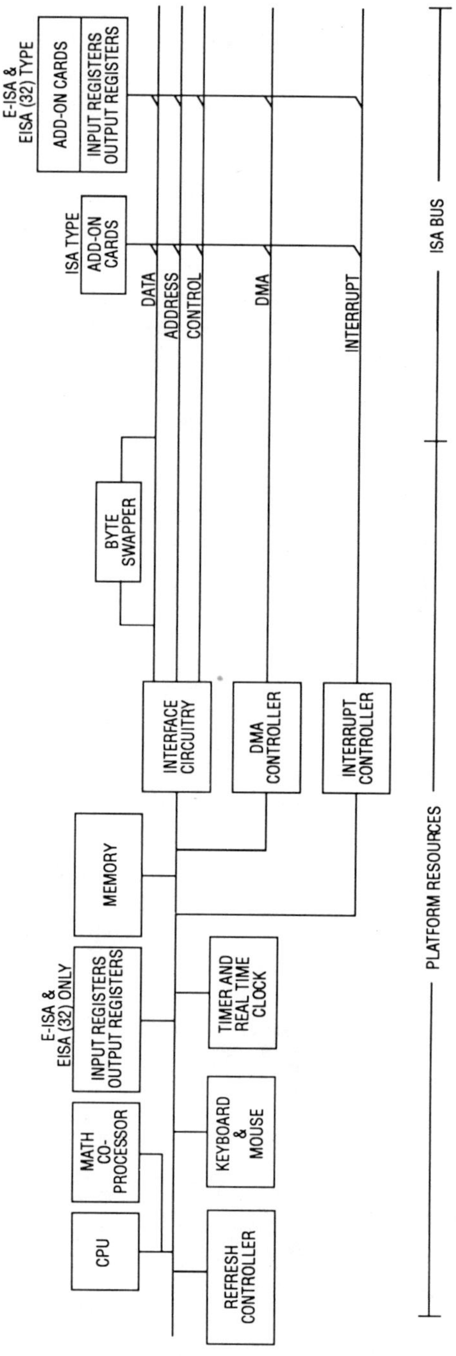

FIGURE 1-1: GENERIC PLATFORM

circuitry exists to interface it correctly to the data width of the bus cycles being executed. The add-on slave card data width must be compatible with the data width of the bus.

DMA CONTROLLER

This is one of the platform resources that can become bus owner. Its purpose is to transfer data between I/O and memory resources with minimum circuitry needed by either resource. It is the only resource on the bus that monitors the DMA signal lines on the bus. An active DMA request signal line will cause the DMA controller to obtain bus ownership and execute transfer cycles.

REFRESH CONTROLLER

This is one of the other platform resources that can become bus owner. Its purpose is to execute refresh cycles for DRAM memory components. It refreshes memory resources both on the platform and on the bus. On the PC and XT it is actually a DMA channel 0 cycle. On the ISA compatible platforms it is separate dedicated circuitry.

INTERRUPT CONTROLLER

This platform resource supports all of the bus interrupt request signal lines. The software can enable or disable individual interrupt channels. It is the responsibility of the interrupt controller to monitor the bus control lines of the enabled channels. An active interrupt signal line on an enabled channel generates an interrupt to the platform CPU, and provides information to identify which interrupt signal lines are active.

KEYBOARD CONTROLLER

All platforms have circuitry to interface to the keyboard. On the PC and XT platforms, this is discrete circuitry. On the AT, this function is provided by the Intel 8042 microcontroller with a built-in BIOS.

TIMER and REAL TIME CLOCK

These platform resources are needed for software compatibility and have no direct relationship with the bus. The timer generates a pulse every 15.6 microseconds to the refresh circuitry to request a refresh cycle.

INTERFACE CIRCUITRY

The purpose of this platform resource is to interface the platform CPU, DMA controller, and refresh controller control signals with the asynchronous control signal lines of the bus. It also contains buffers to interface the platform bus to the ISA bus.

BYTE SWAPPER

The byte swapper circuitry ports bytes between low and high byte data lines on the platform and bus. The low byte data lines contain data bits 0-7. The high byte data lines contain data bits 8-15.

ISA BUS

This is a set of connectors interconnected by the ISA bus signal lines. Each connector site is called a slot. Two types of connectors can reside at each slot: 8 bit and 8/16 bit. The 8 bit slot is a single 62 pin connector, and only supports 8 bit add-on slave cards. The 8/16 bit slot contains both the 8 bit slot 62 pin connector and an additional 36 pin connector. This expanded connector supports a larger address space, additional DMA channels, additional interrupt request lines, and additional data lines. The expanded connector also allows add-on bus owner cards to be used.

ADD-ON CARDS

The bus allows the insertion of add-on cards. Add-on cards that only respond to access, transfer, or refresh cycles are defined as add-on slave cards. Add-on cards that can become bus owners are defined as add-on bus owner cards. An add-on card may be of both types; but, for a given bus cycle, it can only be either a bus owner or a slave.

> *ISA add-on slave cards will work on AT (ISA) platforms. The 8 bit version of these cards will also work on PC and XT platforms, although these platforms execute longer bus cycles.*

> *Add-on bus owner cards that adhere to the ISA platform model and the 8 Mhz timing specifications will operate on 8 Mhz or faster AT (ISA) platforms. An AT platform that operates at bus clock speeds less than 8 Mhz may not have platform memory or I/O resources that will be compatible with a faster add-on bus owner card. The PC and XT do not support add-on bus owner cards.*

E-ISA COMPATIBLE PLATFORM

The lack of an industrial standard for the AT bus, prior to the efforts of Intel, the IEEE P996 committee, and the EISA consortium, has resulted in the defacto ISA "standard". Unfortunately, this defacto "standard" has had to reflect design errors in early platforms, misinterpretation of bus cycles, and other types of errors and inconsistencies. The operation of the 8 and 8/16 bit slots in a platform compatible with the EISA bus specification must maintain compatibility with the ISA standard, at the same time allowing the use of the enhancements outlined in the EISA bus specification. In order to differentiate between 8 and 8/16 bit slots on platforms that adhere to the ISA standard versus platforms and add-on cards that adhere to the full EISA bus specification, the concept of E-ISA compatible platforms and add-on cards has been adopted.

> THE E-ISA COMPATIBLE PLATFORM SHOULD BE VIEWED SIMPLY AS A CONCEPT. IT IS USED TO DISTINGUISH BETWEEN "HISTORICAL" ISA BUS PLATFORMS (AT) AND THE "ENHANCED" ISA BUS (8 AND 8/16 BIT SLOTS CONFORMING TO THE EISA BUS SPECIFICATION).

An E-ISA platform is conceptually the same as the ISA platform. It only supports 8 and 8/16 bit slots, and is compatible with ISA compatible add-on cards. However, it provides enhancements provided by the EISA bus specification that can be exploited by E-ISA compatible add-on cards.

An E-ISA compatible add-on card is conceptually the same as an ISA add-on card. It can function in ISA and E-ISA 8 and 8/16 bit slots, and can also be used in a platform that adheres to the full EISA bus specification.

The significance of an E-ISA compatible platform is that it is an enhanced ISA platform that only adopts the 8 and 8/16 "asynchronous" slots of the EISA bus specifications. It does not adopt the far greater overhead of the full 8/16/32 "synchronous" slots of the EISA bus specifications.

The enhancements of an E-ISA compatible platform or add-on card versus an ISA compatible platform or add-on card are as follows:

- Configuration space is supported.

- I/O space addressing is slot specific.

- The maximum BCLK signal line frequency is increased to 8.33 Mhz.

- Any DMA channel may be programmed for 8 or 16 bit transfers.

- Add-on I/O slave cards can support faster DMA transfer rates.

- The INTERRUPT signal lines can be programmed for edge or level trigger. The level trigger provides the "ORING" of several interrupts on a single INTERRUPT signal line.

- The IOCHRDY and DRQ signal lines are referenced to the BCLK signal line by the platform CPU and DMA controller, respectively.

- The IOCHRDY, SRDY*, and MEMCS16* signal lines have broader support.

- The TC signal line is bi-directional.

> *Except for the Configuration space, the above enhancements do not change the architecture of the platform or add-on cards. The Configuration space consists of input and output registers within the I/O address space which are reserved for board and add-on card configuration. These input and output registers allow software configuration and monitoring of hardware. (See Figure 1-1.)*

The enhancements defined above allow faster DMA transfer rates for E-ISA add-on I/O slave cards, but not for either ISA or E-ISA add-on memory slave cards. This is a result of the EISA bus specification, which defines only the normal ISA compatible DMA transfer between ISA add-on I/O and memory slave cards. The faster DMA transfers are defined between E-ISA add-on I/O slave cards and platform memory or EISA(32) add-on memory cards. The EISA(32) add-on memory cards use the synchronous control lines which are not defined on an E-ISA platform.

The EISA bus specification does not support faster fly-by DMA transfers with an ISA add-on memory card. The reason for this is that the transfer must be done in two cycles, one of which is for the ISA add-on memory card. The ISA add-on memory card cycle is the same as an add-on bus owner card accessing the memory resource. This method slows down the transfer and does not result in a net benefit. See the

TRANSFER CYCLE section of Chapter 3 for more information.

> *E-ISA add-on slave cards will work on ISA platforms. The 8 bit version of these cards will also work on PC and XT platforms, although these platforms execute longer bus cycles. Any software written for the add-on card must be adjusted for the loss of the configuration space.*

> *E-ISA add-on bus owner cards can execute cycles commensurate with an 8.33 Mhz BCLK signal line frequency. On an ISA platform, the E-ISA add-on bus owner card must execute cycles commensurate with an 8 Mhz BCLK signal line frequency. An ISA platform may not have platform memory or I/O resources that will be compatible with a faster add-on bus owner card. The PC and XT do not support add-on bus owner cards.*

EISA(32) COMPATIBLE PLATFORM

The EISA(32) platform is conceptually identical to an ISA platform. Its support of 8 and 8/16 bit slots is the same as the ISA platform with the enhancements supported by an E-ISA platform. The major distinction for EISA(32) is that some slots have a special connector. The design of the connector allows add-on cards for ISA and E-ISA platforms to use the slot. The special connector supplies additional address and data signal lines, additional add-on bus owner card support functions, and synchronous control signal lines.

The synchronous control signal lines provide a new set of capabilities. The ISA and E-ISA buses are essentially asynchronous; that is, most of the signal lines on the 8 and 8/16 bit slots become active and inactive independent of the bus clock. The EISA(32) platform bus still operates the signal lines on the 8 and 8/16 bit slots asynchronously, but some additional EISA(32) cycles unique to EISA are executed using only the synchronous control signal lines on the special connector. The support of both asynchronous and additional synchronous cycles allows EISA(32) platforms to be compatible with ISA and E-ISA compatible add-on cards, and also provides a method for improved performance with the synchronous cycles.

> *Even though the SRDY* signal line on the ISA platform is synchronized to the BCLK signal line, the ISA platform is still defined as an asynchronous bus. Similarly, an E-ISA platform is essentially an asynchronous bus, even though it does synchronize some other bus signal lines.*

> *The existence of slots on an EISA(32) platform that support 8 and 8/16 bit slots with asynchronous cycles and 8/16/32 bit slots with synchronous cycles presents some problems. First, the EISA(32) bus must be able to execute both asynchronous and synchronous access cycles using two sets of control lines. Thus, the EISA(32) platform has EXTENDED INTERFACE CONTROL CIRCUITRY. (See Figure 1-1.) It functions in the same fashion as the INTERFACE CONTROL CIRCUITRY on the ISA and E-ISA platforms; that is, it is the interface between platform resources and the bus's asynchronous control lines. It is also the interface between platform resources and the bus's synchronous control lines. In addition, it provides an interface between EISA(32) or ISA/E-ISA compatible add-on bus owner cards and ISA/E-ISA or EISA(32) compatible add-on slave cards, respectively. Furthermore, for transfer cycles, it provides the interface between the asynchronous and synchronous signal lines. For transfer cycles, it also does data alignment and multiple cycle transfers.*

Tables 1-1 and 1-2 summarize how different add-on cards and platforms interact. For these tables it is assumed that an EISA(32) add-on card by definition uses the synchronous control lines independent of data size. It is also assumed that the ISA and E-ISA platforms use the AT style connector.

PLATFORM TYPE	SLOT SIZE (BITS)	ADD-ON SLAVE CARD TYPE ISA	E-ISA	EISA(32)	CYCLE TYPE
ISA/E-ISA	8	8	8	NONE	ACCESS ASYN
ISA/E-ISA	8/16	8, 16	8, 16	NONE	ACCESS ASYN
ISA/E-ISA	8	8	8	NONE	TRANS. ASYN
ISA/E-ISA	8/16	8, 16	8, 16,	NONE	TRANS. ASYN
EISA(32)	8/16/32	8, 16	8, 16	-	ACCESS ASYN
EISA(32)	8/16/32	8, 16	8, 16	-	TRANS. ASYN
EISA(32)	8/16/32	-	-	8, 16, 32	ACCESS SYN (2)
EISA(32)	8/16/32	-	-	8, 16, 32	TRANS. SYN

TABLE 1-1: PLATFORM AND ADD-ON SLAVE CARD INTERACTION

PLATFORM TYPE	SLOT SIZE (BITS)	ADD-ON BUS OWNR CRD TYPE ISA	E-ISA	EISA(32)	CYCLE TYPE
ISA/E-ISA	8/16	8, 16	8, 16	NONE	ACCESS ASYN
EISA(32)	8/16/32	8, 16	8, 16	-	ACCESS ASYN (1)
EISA(32)	8/16/32	-	-	8, 16, 32	ACCESS SYN (2)

TABLE 1-2: PLATFORM AND ADD-ON BUS OWNER CARD INTERACTION

NOTES:
(1) Accesses to EISA(32) compatible resources by an ISA or E-ISA compatible add-on bus owner card are translated to synchronous cycles by the EXTENDED INTERFACE CONTROL CIRCUITRY.
(2) 8 Bit is actually a combination of SYN. and ASYN.

SUMMARY OF BUS STANDARD DIFFERENCES

Tables 1-3 to 1-7 are summaries of the differences between platforms relative to the platform CPU, add-on cards, and DMA controllers. The other chapters in this book will explain these differences in detail. Because this book focuses on 8 and

8/16 bit slots, the full EISA 8/16/32 (EISA(32)) will not be discussed in any extensive detail.

PLATFORM	IOCHRDY REF. TO BCLK	MAX BCLK (MHZ)	INTERRUPT TRIGGER	CONFIG. SPACE	I/O SLOT ADDRESS
PC	NA	4.77	EDGE	NO	NO
XT	NA	4.77	EDGE	NO	NO
AT	NO	8.00	EDGE	NO	NO
ISA	NO	8.00	EDGE	NO	NO
E-ISA	YES	8.33	EDGE/LVL	YES	YES

TABLE 1-3: GENERAL PLATFORM FEATURES

PLATFORM	NO WAIT 8 BIT	NO WAIT 16 BIT	STANDARD 8 BIT	STANDARD 16 BIT	READY 8 BIT	READY 16 BIT	IOCHRDY REF. TO BCLK
PC	NA	NA	MEM,I/O	NA	MEM,I/O	NA	NO
XT	NA	NA	MEM,I/O	NA	MEM,I/O	NA	NO
AT	NO	MEM	MEM,I/O	MEM,I/O	MEM,I/O	MEM,I/O	NO
ISA	NO	MEM	MEM,I/O	MEM,I/O	MEM,I/O	MEM,I/O	NO
E-ISA	MEM,I/O	MEM	MEM,I/O	MEM,I/O	MEM,I/O	MEM,I/O	YES

TABLE 1-4: PLATFORM CPU CYCLE SUPPORT

PLATFORM	CYCLE TYPE						IOCHRDY REF. TO BCLK
	NO WAIT 8 BIT(1)	NO WAIT 16 BIT	STANDARD 8 BIT(1)	STANDARD 16 BIT	READY 8 BIT(1)	READY 16 BIT	
PC	NA	NA	NA	NA	NA	NA	NO
XT	NA	NA	NA	NA	NA	NA	NO
AT	NO	NO	NO	MEM,I/O	NO	NO	NO
ISA	NO	NO	NO	MEM,I/O	NO	NO	NO
E-ISA	MEM,I/O	MEM	MEM,I/O	MEM,I/O	MEM,I/O	MEM,I/O	YES

TABLE 1-5: ADD-ON BUS OWNER CARD CYCLE SUPPORT

NOTES:

NA = Not applicable
(1) Add-on card begins cycle as a 16 bit access; 8 bit accessed resources force two 8 bit cycles.

PLATFORM	IOCHRDY	SRDY*	MEMCS16*
PC	NA	NA	NA
XT	NA	NA	NA
AT	NO	NO	NO
ISA	YES	YES	YES
E-ISA	YES	YES	YES

TABLE 1-6: ADD-ON BUS OWNER CARD ACCESS TO PLATFORM MEMORY

PLATFORM	TRANSFER TYPES COMP.	"A"	"B"	TC (1)	PROG. CHAN. SIZE	DEMAND & BLOCK TRANSFER INTERRUPTIBLE
PC	YES	NO	NO	UNI	NO	NO
XT	YES	NO	NO	UNI	NO	NO
AT	YES	NO	NO	UNI	NO	NO
ISA	YES	NO	NO	UNI	NO	NO
E-ISA	YES	YES	YES	BI	YES	YES

TABLE 1-7: DMA CONTROLLER

NOTES:

(1) UNI = Unidirectional, BI = Bidirectional

SPECIAL E-ISA PLATFORM NOTE

The enhancements to the portion of the EISA specification that are 8/16 bit ISA compatible, as previously described, cover both hardware, firmware, and software. In terms of hardware, one of the major enhancements is the referencing of the signal lines to the BCLK signal line. Many of these references are related to the extensive use of the BCLK signal line by the full EISA 32 bit bus implementation. For the purposes of this book, the defining of E-ISA is to clearly flag these enhancements over the ISA bus. The enhancements defined relative to the BCLK signal line were chosen for two reasons: first, an ISA platform can make use of these references to better define cycle operation, and, second, the use of these references may allow improved bus performance by shortening cycles. Other EISA enhancements could have been included in the E-ISA definition, but may not have improved either the performance or the understanding of the bus cycles.

CHAPTER TWO

MEMORY AND I/O ADDRESS SPACE

INTRODUCTION

ISA, PC, XT, and E-ISA platforms contain two address spaces: memory and I/O. To maintain compatibility for software written for earlier platforms, there are certain requirements placed on the memory and I/O address spaces. This chapter will briefly outline the requirements relative to add-on card design. For software development, more detailed information is required from the platform developer.

MEMORY ADDRESS SPACE

The memory address space is divided into three sections: platform memory, EPROM, and slot memory resources. The first 640K bytes (00 0000H to 09 FFFFH) is reserved for platform and bus memory resources. (See Figure 2-1.) The next 256K bytes (0A 0000H to 0D FFFFH) are assigned for bus memory resources such as video and hard disk. (Some of these resources may be built on the platform, so not all of the accesses to these address locations will result in bus cycles.) The next 128K bytes (0E 0000H to 0F FFFFH) are assigned to the ROM sockets, which include space for the BIOS.

For a PC or XT compatible platform, 20 ADDRESS signal lines are supported; consequently, the total address space of 1 megabyte is defined from 00 0000H to 0F FFFFH. The ISA and E-ISA platforms support 24 ADDRESS signal lines; consequently, a total address space of 16 megabytes is defined

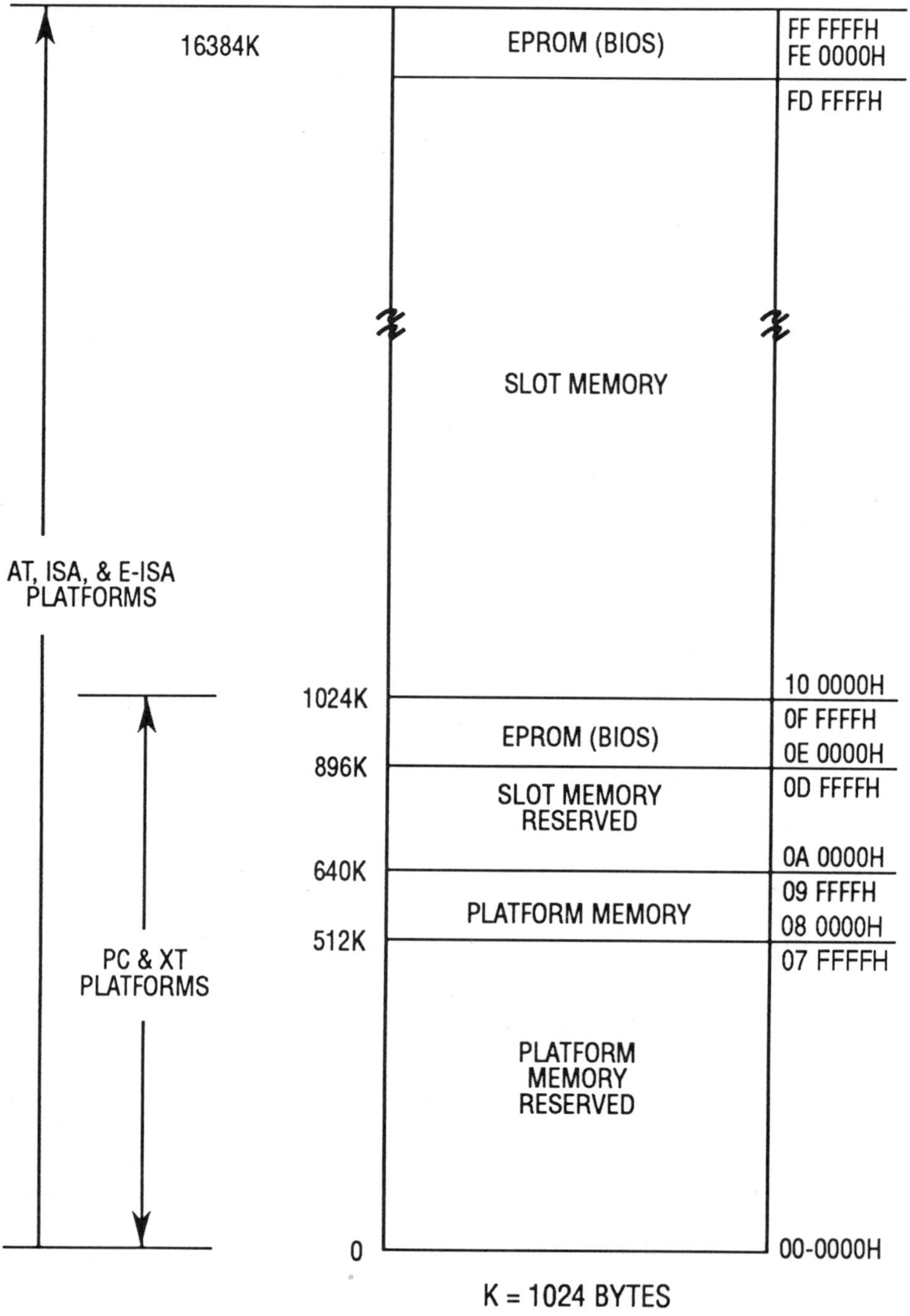

FIGURE 2-1: MEMORY ADDRESS SPACE

from 00 0000H to FF FFFFH. The first 1 megabyte is defined in the same fashion as for the PC and XT. The 128K at the top of the 16 megabyte address space (FE 0000H to FF FFFFH) is a copy of the ROM (BIOS) space in the first megabyte. The remaining memory address space (10 0000H to FD FFFFH) is assigned to bus memory resources.

> *Some platforms are designed with some of the bus memory address space between (10 0000H to FD FFFFH) allocated for platform memory resources. The first 512K of the address space (00 0000H to 07 FFFFH) is designated for interrupt vectors, BIOS data, DOS, and user applications.*

I/O ADDRESS SPACE

The size of the I/O address space of the platform CPUs used on the PC, XT, ISA, or E-ISA platforms is 64K. However, the size of the available I/O address space varies from platform to platform.

On the PC, XT, AT, and ISA platforms, it historically has been assumed that only 1K of the total 64K I/O address space is used. The first 256 bytes are reserved for I/O platform resources, such as the interrupt and DMA controllers (addresses 0000H to 00FFH). (See Figure 2-2.) The remaining 768 bytes are available to "general" I/O slave card resources (addresses 0100H to 03FFH). In that only 1K of the address space was supported, ISA add-on I/O slave cards only decode the first 10 ADDRESS signal lines; consequently, the SA<10-15> signal lines can be any value. The result of not decoding all of the ADDRESS signal lines results in the 256 platform address locations and the 768 bus address locations being repeated 64 times in the address space. (See Figure 2-2.) This repetition occurs on 1K address boundaries.

Add-on I/O slave cards do not use the I/O addresses reserved for the platform resources. Software running on the platform also needs only one of the 64 repetitions of the 256 address locations to access platform resources. The E-ISA and

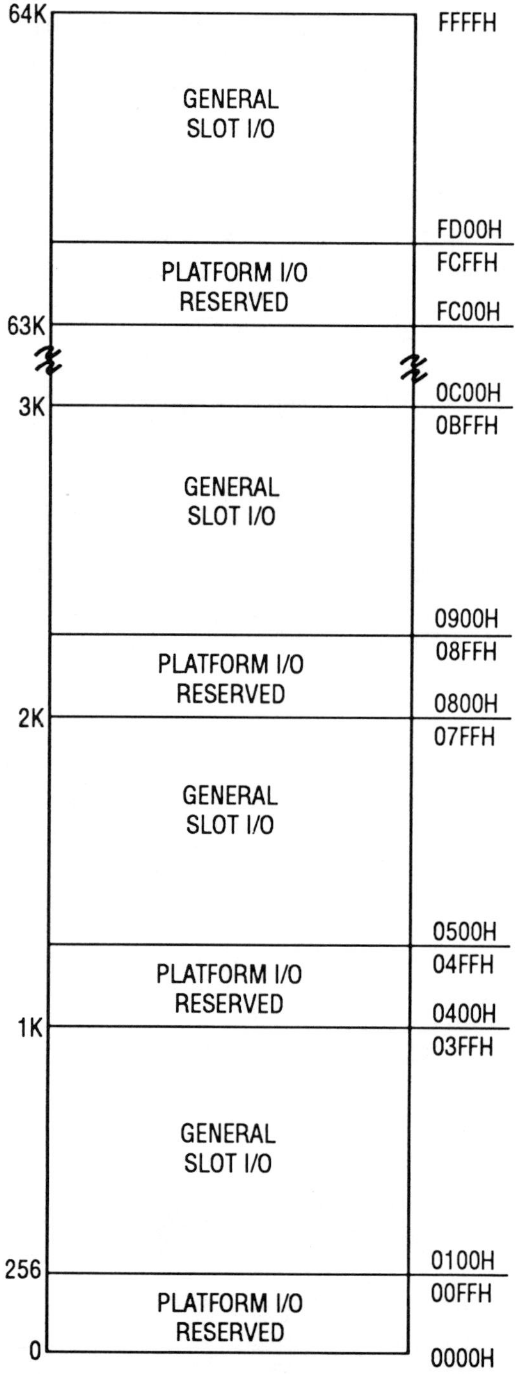

FIGURE 2-2: PC, XT, AT, & ISA ADDRESS SPACE

EISA(32) platforms exploit these facts by assigning 63 of the 64 repetitions of the 256 address locations to specific add-on card slots. The partitioning is such that 4 groups of 256 address locations is assigned to each slot for a total of 1024 specific address locations per slot. (See Figure 2-3.)

> *Slot 0 is defined as the platform; thus, a total of 1K I/O addresses are assigned to the platform. The first 256 address locations are the same platform resources as defined across all platforms. The remaining three groups of 256 address locations are defined for E-ISA and EISA(32) platforms according to the EISA bus specification.*

Since existing add-on slave I/O cards only decode the first ten ADDRESS signal lines, they will respond to the "general" 768 bytes that are repeated 64 times. To maintain compatibility with these boards, this repetition is maintained on all platforms including E-ISA and EISA(32). (See Figure 2-3.)

An add-on card can be installed in any slot. To prevent the requirement for a group of add-on card jumpers to establish a slot address, the E-ISA and EISA(32) platforms have adopted a novel approach. The AEN signal line is used by all platforms to indicate that a DMA transfer cycle is being executed. An active AEN signal line instructs add-on slave I/O card resources to ignore the COMMAND signal lines unless its respective DACK* signal line is active. E-ISA and EISA(32) have an individual AEN signal line attached to each slot. When the access is one of the four groups of 256 address locations of a specific slot, the AEN signal line of that slot is driven inactive. The AEN signal lines of the other slots are driven active. Of course, when the access is to the "general" 768 address locations, all AEN signal lines of all of the slots are driven inactive.

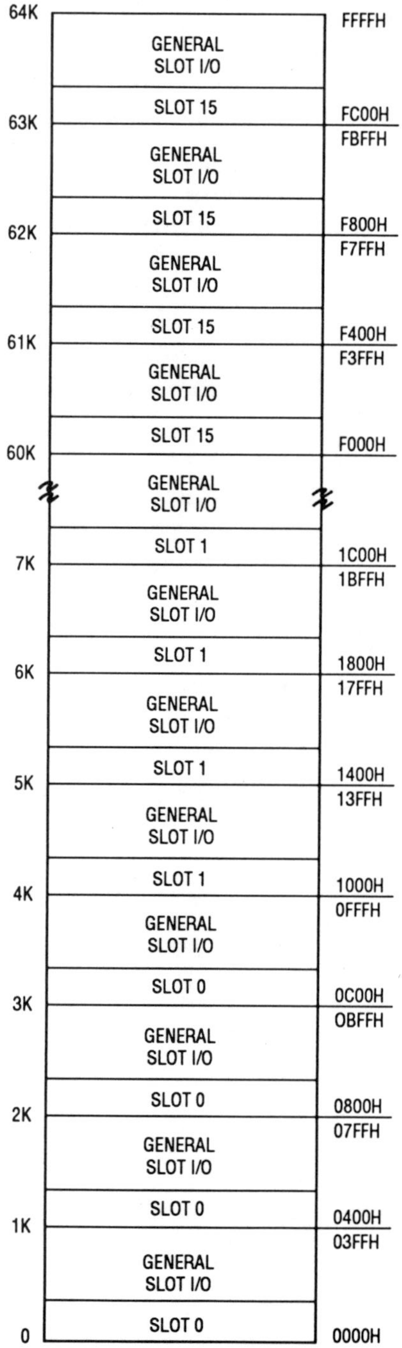

FIGURE 2-3: E-ISA I/O ADDRESS SPACE

> *Even though an E-ISA compatible add-on I/O slave card has access to the SA<0-15> signal lines, only the SA<0-11> signal lines are decoded. The control of the individual AEN signal lines of each slot eliminates the need to decode the SA <12-15> signal lines.*

> *The EISA method of assigning four groups of 256 address locations allows 16 slots to be defined. Traditionally, only 8 to 10 slots have been available on a platform. The AC timings specified by the IEEE P996 and the EISA bus specifications are only for 8 slots.*

CHAPTER THREE

GENERIC BUS CYCLES

INTRODUCTION

This chapter is an introduction to the basic bus cycles. It is important to understand the basic bus cycle concepts in order to understand the information in Chapter 4. A more detailed explanation of the signal lines and bus cycles are in Chapters 5 through 9.

The basic bus cycles on an ISA platform are asynchronous. With one exception, signai lines are not referenced to the system clock. Some signal lines change logic state at any time, while others change within a specific time relative to a logic state change of other signals. The basic bus cycles on an E-ISA platform are also essentially asynchronous, although a few signal lines are referenced to the system clock. Except for a few situations, the system clock is not really used in the traditional manner of a synchronous bus. On either the ISA or E-ISA platforms, the oscillator (OSC) signal line is not used for synchronization at all, and will be discussed in more detail in Chapter 5.

There are four different bus cycles: bus owner access, DMA transfer, memory refresh, and bus arbitration. The bus owner access cycle (access cycle) is executed by the platform CPU or an add-on bus owner card. The DMA transfer cycle (transfer cycle) is executed by the DMA controller. The memory refresh cycle (refresh cycle), is executed by the refresh controller to refresh any volatile memory. The add-on bus owner card arbitration cycle (arbitration cycle), is executed by the add-on bus owner card to obtain bus ownership.

ACCESS CYCLES

The access cycle is executed by the platform CPU or the add-on bus owner card to read and write data with memory and I/O resources. There are three types of access cycles: standard, no wait state, and ready.

The standard access cycle executed when the platform CPU is bus owner is shown in Figure 3-1. The bus address latch enable line, BALE, is pulsed active at the beginning of a valid bus cycle, and indicates that the ADDRESS and associated signal lines on the bus are valid.

The accessed resource on the bus responds to the address with an indication of its size. Either the MEMCS16* or IOCS16* signal line is activated if the resource is 16 bits; otherwise, the resource is assumed to be 8 bits. The COMMAND signal lines, which include SMEMR*, SMEMW*, MEMR*, MEMW*, IOR*, and IOW*, are activated by the bus owner. The length of the active COMMAND signal line is predetermined. Its length depends on the resource data width and whether the cycle is memory or I/O. For a write cycle, the active COMMAND signal line indicates to the accessed resource valid data on the bus. For a read cycle, the active COMMAND signal line provides a reference point to the accessed resource as to when data on the bus must be valid.

The standard access cycle executed by an add-on bus owner card is the same as for the platform CPU, except that the BALE signal line is held active. Consequently, the COMMAND signal lines becoming active are the only indication of the beginning of the cycle. (See Figure 3-2.) The lack of a pulsed BALE signal line also requires that both the LA<17-23> and SA<0-19> signal lines are held valid for the entire active period of the COMMAND signal lines. See Chapter 6 for more detailed information.

The second type of access cycle is the ready access cycle. (See Figures 3-1 and 3-2.) It can be executed by either the platform CPU or by an add-on bus owner card. The standard access cycle becomes a ready access cycle when the accessed resource deactivates the IOCHRDY signal line. The COMMAND signal line will remain active until the IOCHRDY signal line becomes active. The result is a longer active COMMAND pulse than the standard access cycle. The ready access cycle allows access cycles to be lengthened for slower resources.

The increase in the length of a ready cycle is a multiple of one BCLK period. The ISA platform implementation of the IOCHRDY signal line does not reference it to the BCLK line; however, the E-ISA platform implementation does. See Chapter 5 for more detailed information.

If the IOCHRDY signal line is inactive at the same time that the SRDY* signal is active, then the cycle is executed as a ready cycle.

The third type of access cycle is the no wait state access cycle, which can only be executed by the platform CPU in an ISA system, or by the platform CPU or add-on bus master card in an E-ISA system. (See Figure 3-3.) The no wait state access cycle allows access cycles to be shortened for faster resources. A standard access cycle becomes a no wait state access cycle when the accessed resource activates the SRDY* signal line. The result is a shorter active COMMAND pulse than in the standard access cycle. See Chapter 6 for more detailed information.

The original IBM AT samples the SRDY line in such a manner that with an 8 Mhz system clock the SRDY signal line is very difficult to use. However, the ISA implementation samples the SRDY* signal line so that it works with an 8.0 Mhz system clock. Also, the IBM and ISA implementation of this bus cycle is only for 16 bit memory access cycles.*

The E-ISA implementation of the ISA bus has made the implementation easier. It has also expanded the definition of no wait state cycles to include 8 bit memory and I/O access cycles. See Chapter 5 and 6 for more detailed information.

The access cycles can support either 8 or 16 bit cycles in both the memory and I/O address spaces. However, not all platforms and bus owners can support all types of access cycles to all resources. The platform CPU does not require the add-on slave cards to support all access cycles for all data widths. Table 3-1 summarizes the different support levels.

DATA SIZE	PLATFORM	BUS OWNER	STANDARD	READY	NO WAIT STATE
8	ISA	PLAT. CPU	MEM I/O	MEM I/O	-
8	ISA	ADD-ON (1)	MEM I/O	MEM I/O	-
16	ISA	PLAT. CPU	MEM I/O	MEM I/O	MEM
16	ISA	ADD-ON	MEM I/O	MEM I/O	-
8	E-ISA	PLAT. CPU	MEM I/O	MEM I/O	MEM I/O
8	E-ISA	ADD-ON (1)	MEM I/O	MEM I/O	-
16	E-ISA	PLAT. CPU	MEM I/O	MEM I/O	MEM
16	E-ISA	ADD-ON	MEM I/O	MEM I/O	-

TABLE 3-1: ACCESS CYCLE SUPPORT
ISA AND E-ISA COMPATIBLE PLATFORMS

NOTES:

(1) See the NOTE in the ADD-ON BUS OWNER CARD section of Chapter 4.

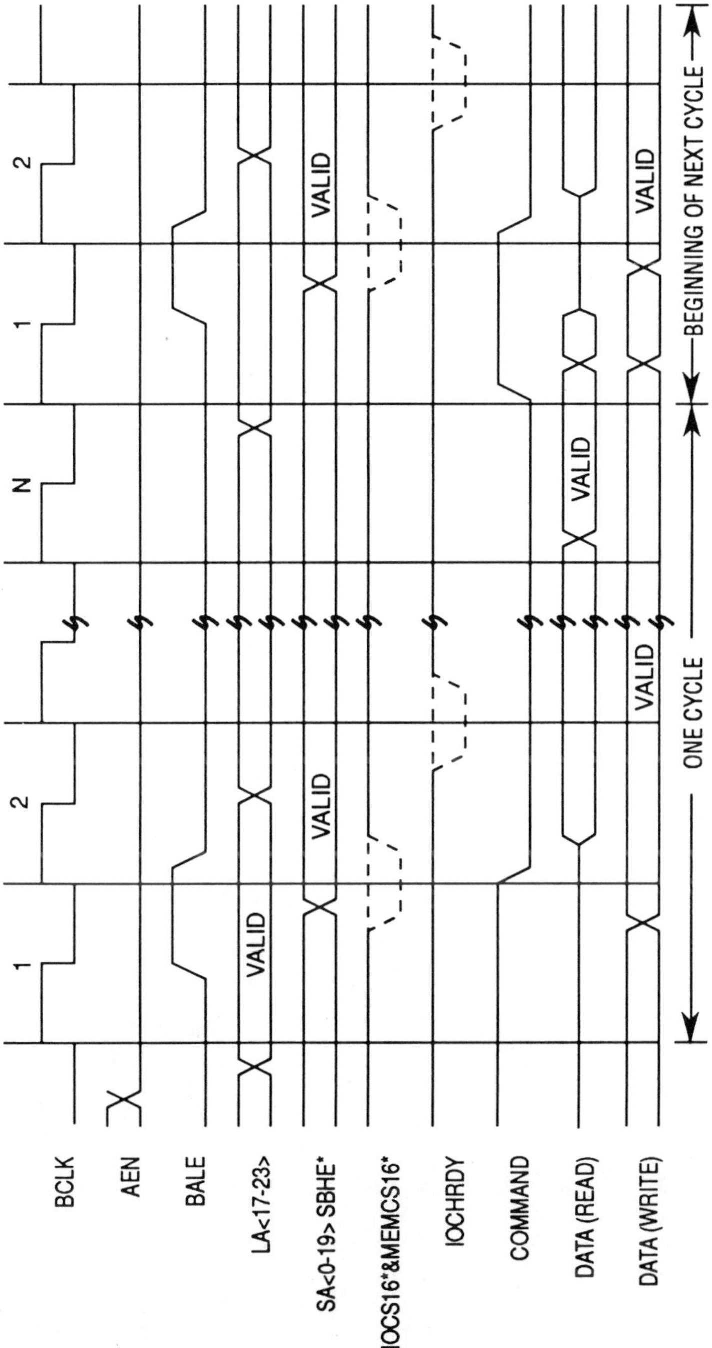

FIGURE 3-1: GENERIC STANDARD AND READY ACCESS CYCLE PLATFORM CPU

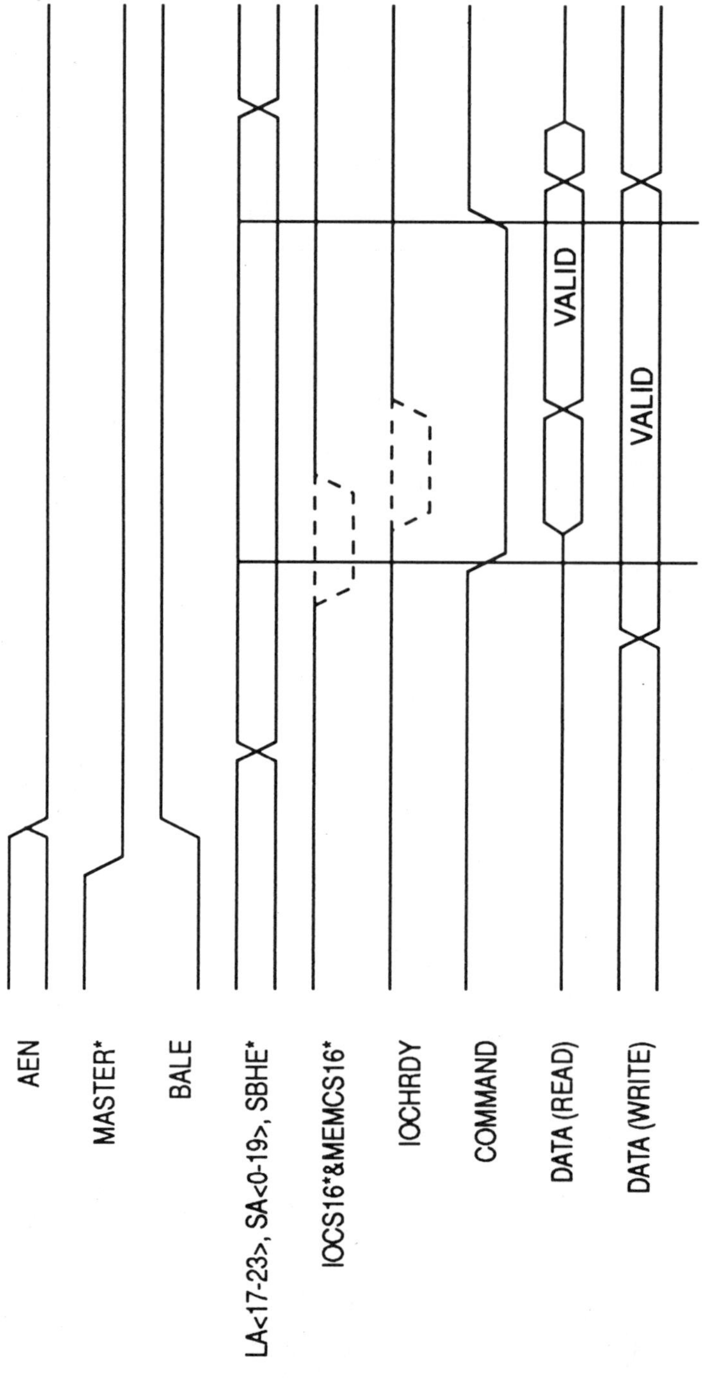

FIGURE 3-2: GENERIC STANDARD AND READY ACCESS CYCLE ADD-ON BUS OWNER CARD

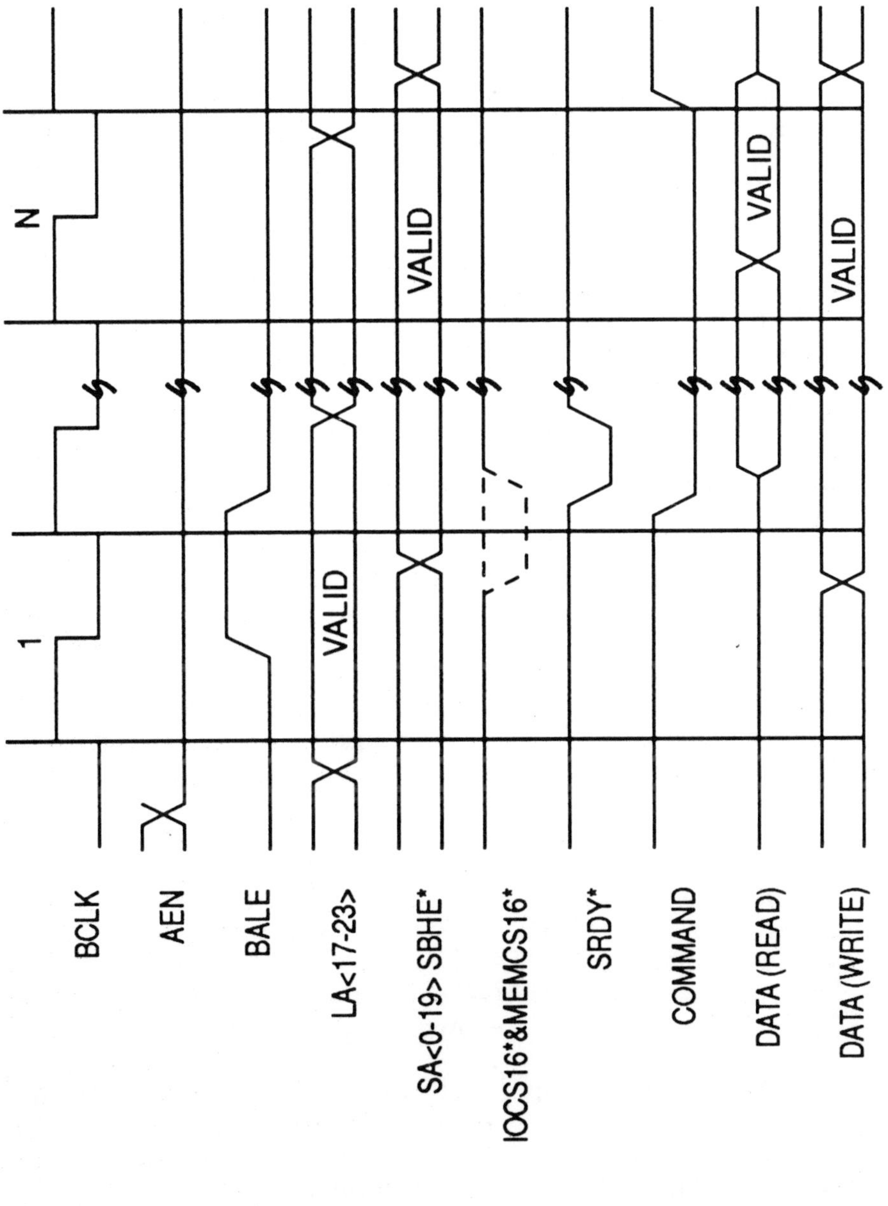

FIGURE 3-3: GENERIC NO WAIT STATE PLATFORM CPU

TRANSFER CYCLES

Transfer cycles operate differently than access cycles. A transfer cycle is executed only by the platform DMA controller to transfer data between I/O and memory resources.

There are two versions of transfer cycles: standard and ready. The execution of either version of the transfer cycle begins with the activation of both a single DMA acknowledge (DACK*) signal line and the AEN signal line by the platform. (See Figures 3-4 and 3-5.) The activation of these lines is in response to an I/O resource activating a DMA request (DRQ) signal line. The activation of the DACK* signal line defines which I/O resource is part of the transfer, and that a transfer cycle is beginning. The AEN line indicates to all other I/O resources that a transfer cycle specific to an I/O resource is occurring, and that the ADDRESS and COMMAND lines must be ignored. Figure 3-4 illustrates the I/O read and memory write transfer cycle, and Figure 3-5 illustrates the I/O write and memory read transfer cycles.

The LA<17-23> and SA<0-19> signal lines are both driven valid prior to the activation of the memory COMMAND signal lines. They are held constant for the entire cycle. The ADDRESS signal lines are only referenced to the memory COMMAND signal lines, because only the memory resources can be addressed. The I/O resources involved in the cycle are "accessed" solely by enabling their specific DACK* signal line.

The length of the COMMAND pulses for the standard transfer cycle is predetermined. A standard transfer cycle becomes a ready transfer cycle if the memory resource deactivates the IOCHRDY within a specific time relative to an active memory COMMAND line. (See Figures 3-4 and 3-5.) The memory COMMAND line will remain active until the IOCHRDY signal line becomes active.

The increase in the length of a ready cycle is a multiple of two BCLK periods on an ISA platform The E-ISA platform extends the ready cycles in multiples of one and two BCLK periods. The ISA platform implementation of the IOCHRDY line does not reference it directly to the BCLK line; however,

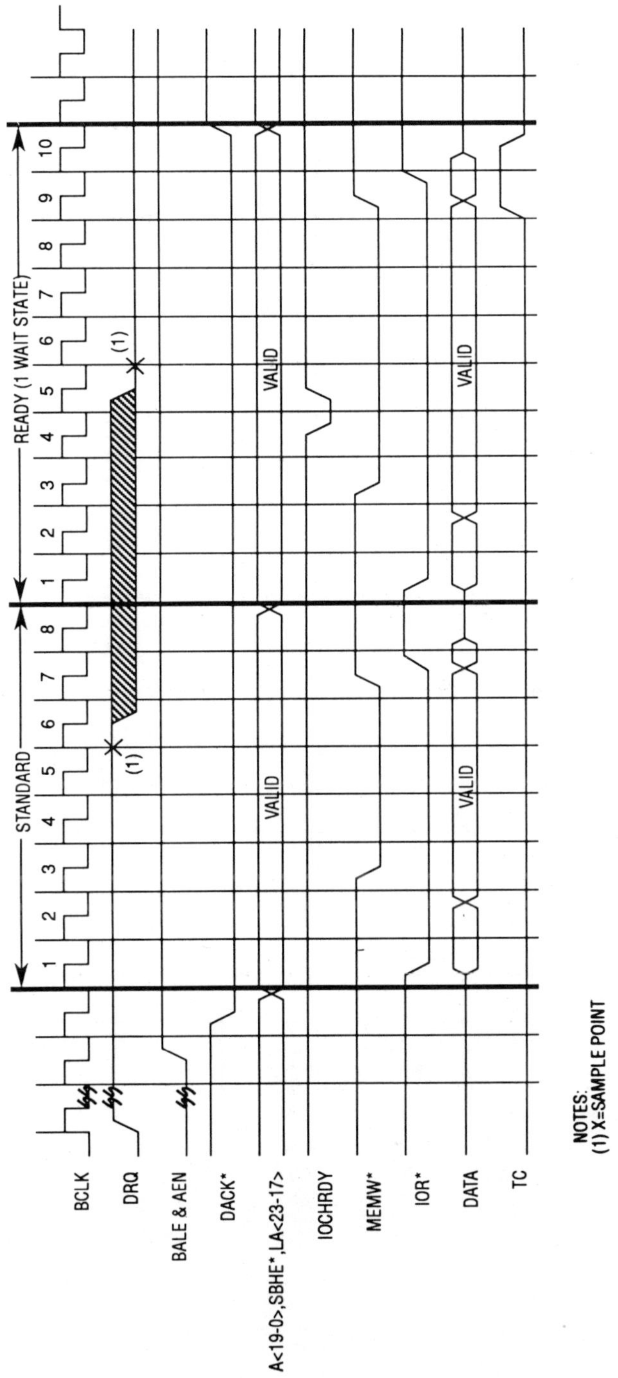

FIGURE 3-4: GENERIC ISA COMPATIBLE TRANSFER CYCLE
I/O READ - MEMORY WRITE

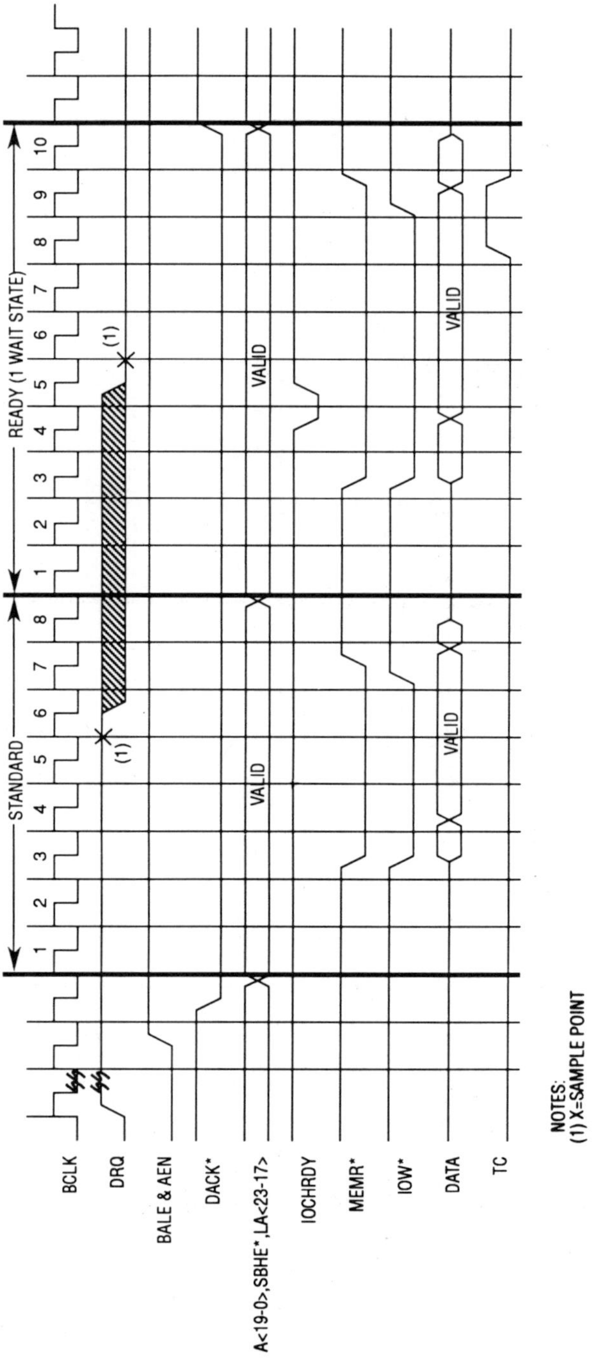

FIGURE 3-5: GENERIC ISA COMPATIBLE TRANSFER CYCLE I/O WRITE - MEMORY READ

the E-ISA platform implementation does. See Chapter 5 for more detailed information.

During a transfer cycle, only the memory resource can inactivate the IOCHRDY signal line. The I/O resource must be ready for the transfer within the cycle length of a standard cycle.

SPECIAL ISA / E-ISA TRANSFER CYCLE ISSUES

On an ISA compatible platform, the bus contains several DMA channels. An add-on slave card can be either the I/O or memory resource in the transfer. Each DMA channel has a pair of DMA request and acknowledge lines. DMA channels numbered 0-3 support only 8 bit transfers, and channels numbered 5-7 support only 16 bit transfers. 16 bit memory resources are designed to allow single bytes to be read and written; consequently, an 8 bit I/O resource can interact with a 16 bit memory resource. The 16 bit I/O resource does not allow single bytes to be read or written. Therefore, transfers between 16 bit I/O and 8 bit memory resources are not supported. These restrictions allow the data to be transferred in a single transfer cycle.

The EISA bus specification has expanded the definition of the transfer cycle, portions of which can be supported on an E-ISA platform. The DMA controller on an E-ISA platform allows each DMA channel to be programmed individually for 8 or 16 bits. In addition to supporting the COMPATIBLE CYCLE, the E-ISA platform can support two additional types of transfer cycles: TYPE "A" and TYPE "B".

The basic operation of the COMPATIBLE cycle is the same as on the ISA platform. (See Figures 3-4 and 3-5.) The operation of the TYPE "A" and TYPE "B" transfer cycles are the same as the COMPATIBLE cycles with two exceptions. First, the cycle length and associated internal cycle timings have been shortened. Second, the basic cycle operation for add-on I/O slave cards is the same as for the COMPATIBLE transfer cycle. (See Figures 3-4 and 3-5.) Table 3-2 summarizes the transfers that take place between different resources. See Chapter 7 for more detailed information.

ADD-ON CARD IO RESOURCE		MEMORY RESOURCE			PLATFORM TYPE	TRANSFER TYPE	FLY -BY
COMPAT	SIZE	COMPAT	LOCATION	SIZE			
ISA E-ISA	8	ISA	PLATFORM	8 16 32	ISA	COMPATIBLE	YES
ISA E-ISA	16	ISA	PLATFORM	16 32	ISA	COMPATIBLE	YES
E-ISA	8	E-ISA	PLATFORM	8 16 32	E-ISA	ALL	YES
E-ISA	16	E-ISA	PLATFORM	16 32	E-ISA	ALL	YES
ISA E-ISA	8	ISA	ADD-ON	8 16	ISA/E-ISA	COMPATIBLE	YES
ISA E-ISA	16	ISA	ADD-ON	16-	ISA/E-ISA	COMPATIBLE	YES

TABLE 3-2: TRANSFER CYCLES

NOTES:

(1): The portion of the transfer cycle to the add-on memory card appears as an ISA add-on bus owner card cycle.

REFRESH CYCLE

The refresh cycle is executed only by the platform refresh controller. There are two versions of the refresh cycle: standard and ready. The execution of either version of the refresh cycle begins with the activation of the REF* signal line. The REF* signal line is driven active by the add-on card when it is bus owner or by the refresh controller. The actual refresh cycle is a simplified memory access cycle. (See Figure 3-6.) The platform or add-on bus owner card will tri-state the ADDRESS and COMMAND signal lines prior to the refresh controller driving them. On an ISA platform, only the SA<0-7> signal lines contain the valid refresh address. On the E-ISA platform, both the LA<17-23> and SA<0-19> signal lines have valid refresh address information. The ADDRESS signal lines are driven valid before the MEMR* and SMEMR* signal lines are driven active. The ADDRESS signal lines are held valid until the COMMAND signal lines are driven inactive.

The length of the standard refresh cycle is predetermined. A standard refresh cycle becomes a ready refresh cycle if the memory resource of the transfer deactivates the IOCHRDY signal line within a specific time relative to active MEMR* signal line. This signal line will remain active until the IOCHRDY signal line becomes active. During a refresh cycle, only the memory resource can inactivate the IOCHRDY signal line.

The increase in the length of a ready cycle is a multiple of one BCLK period. The ISA platform implementation of the IOCHRDY signal line does not reference it to the BCLK signal line, but the E-ISA platform implementation does. See Chapter 5 for more detailed information.

ARBITRATION CYCLE

The arbitration cycle is the method by which an add-on bus owner card can become the bus owner. The cycle actually begins as a DMA transfer cycle. The add-on bus owner card drives one of the DRQ signal lines active. (See Figure 3-7.) The DMA controller responds by driving the associated DACK* signal line active. The add-on bus owner card has a finite amount of time to drive the MASTER* signal line active to obtain ownership of the bus. Once the DMA controller has determined that it is not a normal transfer cycle, it tri-states the ADDRESS, COMMAND, and DATA signal lines. The add-on bus owner card can proceed to execute access cycles to memory and I/O resources. The add-on bus owner card relinquishes bus ownership by driving the DRQ and MASTER* signal lines inactive. See Chapter 9 for more detailed information.

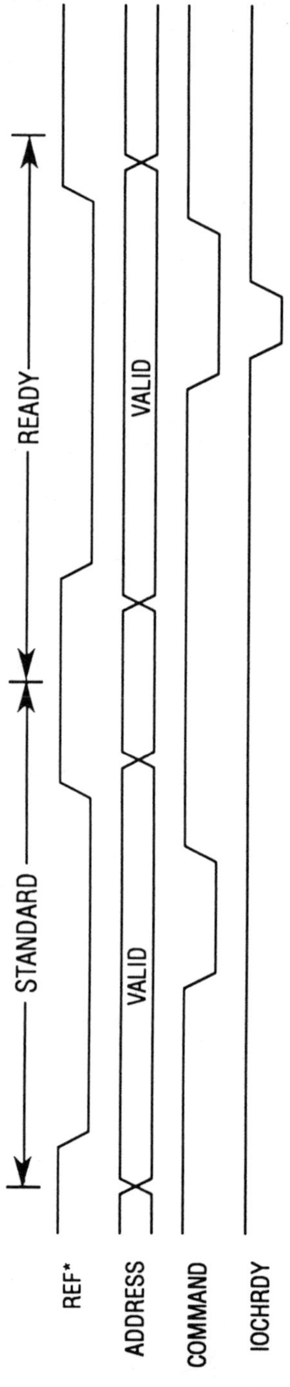

FIGURE 3-6: REFRESH CYCLE

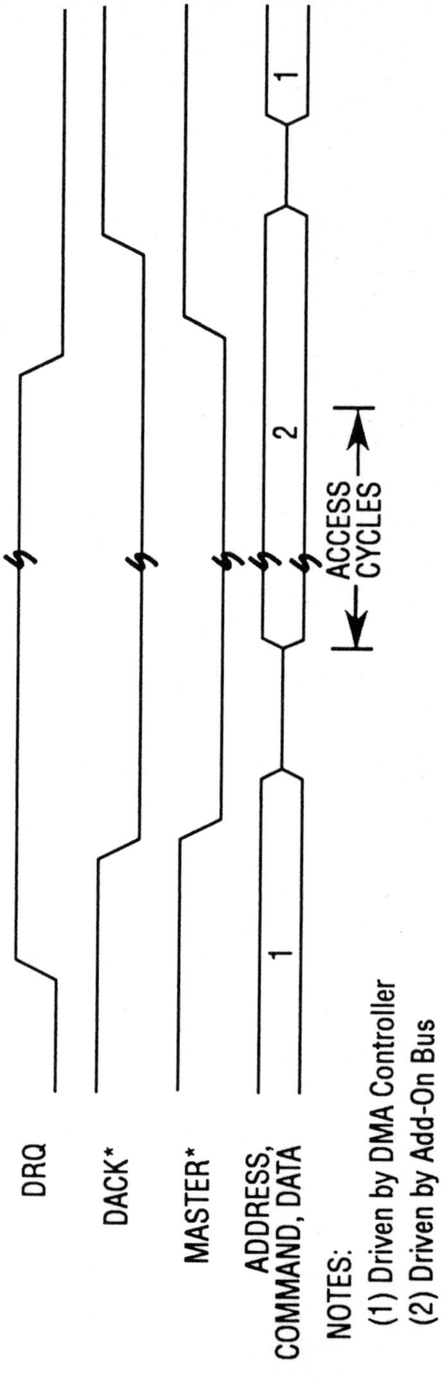

FIGURE 3-7: ARBITRATION CYCLE

CHAPTER FOUR

INTERACTION OF ISA AND E-ISA ADD-ON CARDS

INTRODUCTION

An add-on card can operate in two basic modes on the bus: either as a slave or as a bus owner. An add-on card can be designed to support either or both modes.

ADD-ON SLAVE CARD

An add-on slave card can support bus access and transfer cycles as outlined in Table 1-1. Refresh and arbitration cycles are also supported. ISA or E-ISA add-on slave cards only interact with the signal lines of 8 and 8/16 bit slots of the ISA and E-ISA bus.

MEMORY CYCLES

An add-on slave card can support memory cycles as an 8 or 16 bit resource. As an 8 bit resource it can reside in either an 8 bit or an 8/16 bit slot. As a 16 bit resource, the add-on slave card can only reside in an 8/16 bit slot.

8 BIT MEMORY RESOURCE

An 8 bit add-on slave card installed in an 8 bit slot only has the first 20 address lines, SMEMR*, and SMEMW* available. Therefore, the memory cycle is restricted to the first 1 megabyte of the address space. An 8 bit resource installed in a

8/16 bit slot has access to all 24 address lines. Consequently, it can reside anywhere within the 16 megabyte address space. If it wants to respond beyond the first 1 megabyte, it must use the MEMR* and MEMW* signal lines and properly decode all of the ADDRESS signal lines.

> *The SRDY* signal line is available on an 8 bit slot. On ISA platforms, a no wait state cycle is restricted to 16 bit memory resources; consequently, only 8/16 bit slots support this cycle. E-ISA platforms support no wait state cycles for 8 bit memory and I/O resources in addition to 16 bit memory resources; consequently, both 8 and 8/16 bit slots support this cycle.*

16 BIT MEMORY RESOURCE

An add-on slave card that is designed as a 16 bit memory resource can only reside in an 8/16 bit slot. Consequently, it can support a larger address space than its 8 bit counterpart installed in an 8 bit slot. All 24 address lines are available to support the full 16 megabyte address space. On an ISA or E-ISA platform, a 16 bit memory resource also supports the MEMR*, MEMW*, and SRDY* signal lines. The support of these lines allows no wait state, standard, and ready cycles to be executed.

An add-on slave card that is designed as a 16 bit memory resource must support the MEMCS16* signal line. The activation of this line indicates to the bus owner that a 16 bit cycle can be executed. The only method that can be used to determine a 16 bit cycle is the decode of the LA<17-23> signal lines by the add-on slave card. The add-on slave card will enable the MEMCS16* signal line if the LA<17-23> signal lines indicate a proper address range for the card.

> *The MEMCS16* signal line is monitored by platform CPUs and add-on bus owner cards during access cycles, as well as by the byte swapper circuit during transfer cycles. The MEMCS16* signal line is not monitored by the DMA controller during transfer cycles, or by the refresh controller during refresh cycles.*

> *The add-on slave card cannot use the memory COMMAND signal lines to qualify a 16 bit access. It must blindly decode the LA<17-23> signal lines and activate the MEMCS16* signal line if required. Consequently, the entire 128K byte block selected by LA<17-23> must be the same data width. This "blind" decode does not use the memory or I/O COMMAND signal lines as qualifiers; consequently, it may activate the MEMCS16* signal line during I/O cycles. It is the responsibility of the bus owner to ignore the MEMCS16* signal line during I/O cycles.*

> *A 16 bit resource that activates the MEMCS16* signal line or supports memory above the first 1 megabyte must use the MEMR* or MEMW* signal lines, and not the SMEMR* or SMEMW* signal lines. The latter two command lines are only active for the first 1 megabyte of the address space.*

ADD-ON MEMORY SLAVE CARD ISSUES

> *The use of the SRDY* signal line allows a shorter cycle time. Some ISA compatible platform CPUs do not support no wait state access cycles. The add-on slave card must be designed to support the longer standard or ready cycles even if it activates the SRDY* signal line. All E-ISA compatible platform CPUs do support no wait state cycles. ISA compatible add-on bus owner cards do not support the SRDY* signal line during access cycles, but E-ISA compatible add-on bus owner cards do.*

The use of the IOCHRDY signal line by the add-on slave card allows the cycle to be extended. Some ISA compatible add-on bus owner cards do not recognize the IOCHRDY signal line, and all of the memory cycles are run as standard cycles. An add-on slave card that activates the IOCHRDY line will expect an extended cycle; consequently, the data may not be valid at the end of the standard cycle. Add-on slave cards should support standard cycles when being accessed by an ISA compatible add-on bus owner card. When an E-ISA compatible add-on bus owner card is the bus owner, the IOCHRDY signal line is supported.

Table 4-1 summarizes which signal lines are driven by which resources during a memory access cycle to an add-on slave card. "PLAT. CIRCUIT" in some cases may simply be a pull-up resistor. In this table, "PLAT." is the abbreviation for "platform".

LINE	DRIVEN BY	LINE	DRIVEN BY
AEN	PLAT. CIRCUIT	MEMR*	BUS OWNER
BALE	PLAT. CIRCUIT	MEMW*	BUS OWNER
DACK*	PLAT. DMA	OSC	PLAT. CIRCUIT
DRQ	I/O RESOURCE	REF*	PLAT. REF. CONT.
IOCHCK*	ANY RESOURCE	RESET	PLAT. CIRCUIT
IOCHRDY	ADD-ON SLAVE	SA	BUS OWNER
IOCS16*	I/O RESOURCE	SBHE*	BUS OWNER
IOR*	BUS OWNER	SD	BUS OWNER (WR)
IOW*	BUS OWNER	SD	ADD-ON SLAVE (RD)
IRQ	ANY RESOURCE	SMR/SMW*	PLAT. CIRCUIT
LA	BUS OWNER	SYSCLK	PLAT. CIRCUIT
MASTER*	ADD-ON BUS OWN	TC	PLAT. DMA
MEMCS16*	ADD-ON SLAVE	SRDY*	ADD-ON SLAVE

TABLE 4-1: ADD-ON SLAVE CARD AS MEMORY RESOURCE

I/O CYCLES

An add-on slave card can support I/O cycles as an 8 or 16 bit resource. As an 8 bit resource, it can only reside in an 8/16 bit slot.

8 BIT I/O RESOURCE

An 8 bit resource installed in either an 8 bit or 8/16 bit slot has access to the first 16 address lines, plus IOR* and IOW* lines. The entire 64K address space is supported.

The SRDY line is available on 8 bit slots. On ISA platforms, no wait state cycles are restricted to 16 bit memory resources. E-ISA platforms support no wait state cycles to 8 bit I/O resources in addition to 8 and 16 bit memory resources.*

16 BIT I/O RESOURCE

A 16 bit I/O resource can only be installed in an 8/16 bit slot. It interacts with the bus in nearly the same fashion as an 8 bit resource, since the ADDRESS and COMMAND signal lines are the same in the two cases. The only difference is in the length of the cycle and the support of the IOCS16* signal line.

An add-on slave card that is designed as a 16 bit I/O resource must support the IOCS16* signal line. The activation of this line instructs the bus owner to execute a 16 bit cycle; otherwise, an 8 bit cycle is executed. The only method that can be used to determine a 16 bit cycle is the decode of the SA<0-15> address lines by the add-on slave card. Because the entire I/O address space can be decoded down to the byte level, there are no restrictions on the grouping together of identical byte widths, as there are for memory resources.

The IOCS16 signal line is monitored by platform CPUs or add-on bus owner cards during access cycles. The IOCS16* signal line is not monitored by the DMA controller during transfer cycles, or by the refresh controller during refresh cycles.*

The add-on slave card must decode the SA<0-9> (ISA) or SA<0-11> (E-ISA) signal lines, and blindly activate the IOCS16 line when appropriate. It cannot use any of the memory or I/O COMMAND signal lines as qualifiers. Consequently, it will activate the IOCS16* signal line during memory cycles. It is the responsibility of the bus owner to ignore the IOCS16* signal line during memory cycles.*

16 bit I/O resources do not support the SRDY signal line. The I/O command for a 16 bit access is activated later relative to the memory and 8 bit I/O commands; consequently, there is insufficient time for the add-on slave card to properly drive the SRDY* signal line.*

ADD-ON I/O SLAVE CARD ISSUES

The use of the IOCHRDY signal line by the add-on slave card allows the cycle to be extended. Some ISA compatible add-on bus owner cards do not recognize the IOCHRDY line, and all of the I/O cycles are run as standard cycles. However, the add-on slave card that activated the IOCHRDY line will expect an extended cycle. Consequently, the data may not be valid at the end of the standard cycle. Add-on slave cards should support standard cycles when being accessed by an ISA compatible add-on bus owner card. When an E-ISA compatible add-on bus owner card is the bus owner, the IOCHRDY signal line is supported.

Table 4-2 summarizes which signal lines are driven by which resources during an I/O access cycle to an add-on slave card. "PLAT. CIRCUIT" in some cases may simply be a pull-up resistor. In this table, "PLAT" is the abbreviation for "platform".

LINE	DRIVEN BY	LINE	DRIVEN BY
AEN	PLAT. CIRCUIT	MEMR*	BUS OWNER
BALE	PLAT. CIRCUIT	MEMW*	BUS OWNER
DACK*	PLAT. DMA	OSC	PLAT. CIRCUIT
DRQ	I/O RESOURCE	REF*	PLAT. REF. CONT.
IOCHCK*	ALL RESOURCES	RESET	PLAT. CIRCUIT
IOCHRDY	ADD-ON SLAVE	SA	BUS OWNER
IOCS16*	ADD-ON SLAVE	SBHE*	BUS OWNER
IOR*	BUS OWNER	SD	BUS OWNER (WR)
IOW*	BUS OWNER	SD	ADD-ON SLAVE
IRQ	ANY RESOURCE	SMR/SMW*	PLAT. CIRCUIT
LA	BUS OWNER	BCLK	PLAT. CIRCUIT
MASTER*	ADD-ON BUS OWN	TC	PLAT. DMA
MEMCS16*	MEM. RESOURCE	SRDY*	PLAT. CIRCUIT (1)

TABLE 4-2: ADD-ON SLAVE CARD AS I/O RESOURCE

NOTE: (1) An E-ISA compatible add-on slave card can drive his signal for an 8 bit I/O access cycle.

ACCESS CYCLE DATA SIZE MISMATCH

An add-on slave card that is a 16 bit memory resource cannot force the bus owner to execute a 16 bit access if the bus owner only intends to execute an 8 bit access cycle. Conversely, a bus owner that intends to execute a single 16 bit accesss cycle cannot do so if the memory resource is only an 8 bit resource. The proper interpretation of the MEMCS16*, SA0, and SBHE* signal lines allows data size mismatch to occur on the ISA or E-ISA platforms.

If the SA0 line is a logic low and the SBHE* signal line is active, the bus owner intends to execute a single 16 bit access. The activation of the MEMCS16* signal line by the memory resource indicates to the bus owner that the access can be completed in the present cycle, since the SD<0-15> signal lines map directly to an even address memory word location. The non-activation of the MEMCS16* signal line by the memory resource forces the access to be completed in two 8 bit cycles. The access will be completed as an 8 bit cycle with the SD<0-7> signal lines mapped directly to an even address memory byte location. The next cycle that the bus owner executes will be an 8 bit access in order to access the upper order byte of the word in the previous cycle. The bus owner will drive the SA0 signal line to a logic high and the SBHE* signal line active to indicate an 8 bit cycle with the data byte on the SD<7-15> signal lines. The data byte has been mapped between the upper order and the lower order data lines by the platform byte swapping circuit. The SD<0-7> signal lines must be mapped directly to the odd address memory byte.

If the SA0 signal line is a logic low and SBHE* signal line is inactive, the bus owner intends to execute a single 8 bit access to an even address byte. The data byte will be on the SD<0-7> signal lines, and can be directly mapped to the even address memory byte. The state of the MEMCS16* signal line does not affect the cycle, because the bus owner will complete the access as an 8 bit cycle.

If the SA0 signal line is a logic high and the SBHE* signal line is active, the bus owner intends to execute a single 8 bit access to an odd address byte with the data on the upper data lines. The non-activation of the MEMCS16* signal line by the memory resource causes the data byte to be mapped between the upper order and the lower order data lines by the platform byte swapping circuit. The SD<0-7> signal lines must be mapped directly to the odd address memory byte. The activation of the MEMCS16* signal line by the memory resource allows the upper order data byte on the SD<8-15> signal lines to be mapped directly to the odd address memory byte.

The operation of an I/O access involving the SA0, SBHE*, and SD<0-15> signal lines is the same as for a memory access.

> *If one of the 8 bit access cycles is a read from an add-on slave card that can drive the upper order data lines, and the platform byte swapping circuitry is active, the upper order data lines must not be driven by the add-on slave card. This prevents buffer fights on the upper order data lines between the byte swapper circuitry and the add-on slave card.*

Table 4-3 outlines the interpretation by the platform byte swapper circuitry of specific signal lines.

BUS OWNER VALID DATA	SBHE*	SA0	ADD-ON SLAVE DATA	XXCS16*	BYTE SWAP	AT ADD-ON CARD VALID DATA	MAPPING ADDRESS
8 BITS							
SD<0-7>	INACT.	0	8	INACT.	NO	SD<0-7>	EVEN BYTE
SD<8-15>	ACTIVE	1	8	INACT.	YES	SD<0-7>	ODD BYTE
SD<0-7>	INACT.	0	16	ACTIVE	NO	SD<0-7>	EVEN BYTE
SD<8-15>	ACTIVE	1	16	ACTIVE	NO	SD<8-15>	ODD BYTE
16 BITS							
SD<0-15>	ACTIVE	0	8	INACT.	NO	SD<0-7>	EVEN BYTE
SD<0-15>	ACTIVE	0	16	ACTIVE	NO	SD<0-15>	EVEN WORD

TABLE 4-3: BYTE SWAPPING FOR ACCESS CYCLES TO ADD-ON SLAVE CARDS

> *The condition of SBHE* inactive and SA0 logical 1 is not allowed.*

SPECIAL ISA/E-ISA ACCESS CYCLE I/O ISSUES

As outlined in Chapter 2, certain I/O address spaces are reserved. The need to reserve these addresses can be traced back to the history of I/O add-on slave card design. Even though the original PC supplied all 16 address lines to the bus, early add-on slave cards only decoded the first 10 address lines. The decode of only the first 10 address lines caused duplicate data to be read or written at every 1K increment from the base address. The first 256 byte block was reserved for platform I/O resources; thus, this block is repeated 63 additional times in the 64K address space. E-ISA platforms exploit this fact by assigning the first four blocks to the platform I/O resources. The next four blocks are assigned to slot 1, the next four blocks are assigned to slot 2, and so forth. This results in 1K bytes of addresses being uniquely assigned to each slot.

For add-on slave cards that decode only the first 10 address lines, any I/O access in the address range of 0100H to 03FFH is valid. The data written and read to this 768 byte block equals those at the same address displaced by 1K multiples. For those add-on slave cards that want to be compatible to ISA-only platforms, the first 10 address signal lines (SA<0-9>) must be decoded, and only the addresses within the 768 byte block will be valid. All other addresses will not constitute valid I/O bus accesses.

The previously described ISA-only compatible add-on card will also work with E-ISA compatible platforms. However, it will not support the E-ISA configuration space and slot specific I/O addressing. To support these special E-ISA functions, the first twelve signal lines (SA<0-11>) must be decoded by the add-on slave card. See Chapters 2 and 12 for more information on the configuration space and slot-specific I/O addresses.

TRANSFER CYCLES

An add-on slave card can request the DMA controller on the platform to transfer data between memory and I/O resources. The add-on slave I/O card requests the transfer by activating one of the DRQ signal lines. The DMA controller acknowledges that the ensuing cycles are DMA cycles that involve the add-on slave I/O card by activating the appropriate DACK* signal line. The memory resource of the transfer may be on the platform or on another add-on slave card.

One difference between a platform CPU access cycle and a DMA transfer cycle is that the I/O and memory COMMAND signal lines are active simultaneously for the DMA transfer cycle. The simultaneous activation of the I/O and memory COMMAND signal lines allows the data to be placed onto and retrieved from the bus during the same cycle. During a DMA cycle, the address on the bus is for memory and does not represent a valid I/O address. The AEN signal line is activated by the platform as indication to the add-on slave I/O cards to ignore the ADDRESS and the I/O COMMAND signal lines. Only the add-on slave I/O card that has activated the DRQ and DACK* signal lines responds to the activation of the I/O COMMAND signal lines. If the transfer includes memory on an add-on memory slave card, only that card responds to the ADDRESS and COMMAND signal lines.

> *The add-on slave I/O cards can distinguish between a valid I/O cycle and a DMA transfer by monitoring all of the DACK* lines. However, this would require an 8 bit resource to reside in an 8/16 bit slot. IBM simplified the design requirements by implementing the AEN signal line approach. See Chapters 5 and 7 for more detailed information.*

Another difference between a platform CPU access and a DMA transfer cycle is the operation of BALE and the LA<17-23> signal lines. BALE is driven to logic 1 by platform circuitry. Consequently, the LA<17-23> signal lines must be driven valid by the DMA controller for the entire transfer cycle.

Tables 4-4 to 4-7 outline which signal lines are driven by which resources under the four transfer structures. "PLAT. CIRCUIT" in some cases may simply be a pull-up resistor.

LINE	DRIVEN BY	LINE	DRIVEN BY
AEN	PLAT. CIRCUIT	MEMR*	PLAT. DMA
BALE	PLAT. CIRCUIT	MEMW*	PLAT. DMA
DACK*	PLAT. DMA	OSC	PLAT. CIRCUIT
DRQ	ADD-ON SLAVE	REF*	PLAT. REF. CONT.
IOCHCK*	ALL RESOURCES	RESET	PLAT. CIRCUIT
IOCHRDY	MEMORY RESOURCE	SA	PLAT. DMA
IOCS16*	PLAT. CIRCUIT	SBHE*	PLAT. DMA
IOR*	PLAT. DMA	SD	MEM. RES. (RD)
IOW*	PLAT. DMA	SD	ADD-ON SLAVE (WR)
IRQ	ANY RESOURCE	SMR/SMW*	PLAT. CIRCUIT
LA	PLAT. DMA	BCLK	PLAT. CIRCUIT
MASTER*	PLAT. CIRCUIT	TC	PLAT. DMA (1)
MEMCS16*	PLAT. CIRCUIT	SRDY*	PLAT. CIRCUIT

TABLE 4-4: ADD-ON SLAVE CARD AS I/O RESOURCE DATA DESTINATION FOR TRANSFER CYCLES

LINE	DRIVEN BY	LINE	DRIVEN BY
AEN	PLAT. CIRCUIT	MEMR*	PLAT. DMA
BALE	PLAT. CIRCUIT	MEMW*	PLAT. DMA
DACK*	PLAT. DMA	OSC	PLAT. CIRCUIT
DRQ	ADD-ON SLAVE	REF*	PLAT. REF. CONT.
IOCHCK*	ALL RESOURCES	RESET	PLAT. CIRCUIT
IOCHRDY	ADD-ON SLAVE	SA	PLAT. DMA
IOCS16*	PLAT. CIRCUIT	SBHE*	PLAT. DMA
IOR*	PLAT. DMA	SD	MEM. RES. (WR)
IOW*	PLAT. DMA	SD	ADD-ON SLAVE (RD)
IRQ	ANY RESOURCE	SMR/SMW*	PLAT. CIRCUIT
LA	PLAT. DMA	BCLK	PLAT. CIRCUIT
MASTER*	PLAT. CIRCUIT	TC	PLAT. DMA (1)
MEMCS16*	PLAT. CIRCUIT	SRDY*	PLAT. CIRCUIT

TABLE 4-5: ADD-ON SLAVE CARD AS I/O RESOURCE DATA SOURCE FOR TRANSFER CYCLES

LINE	DRIVEN BY	LINE	DRIVEN BY
AEN	PLAT. CIRCUIT	MEMR*	PLAT. DMA
BALE	PLAT. CIRCUIT	MEMW*	PLAT. DMA
DACK*	PLAT. DMA	OSC	PLAT. CIRCUIT
DRQ	I/O RESOURCE	REF*	PLAT. REF. CONT.
IOCHCK*	ALL RESOURCES	RESET	PLAT. CIRCUIT
IOCHRDY	I/O RESOURCE	SA	PLAT. DMA
IOCS16*	PLAT. CIRCUIT	SBHE*	PLAT. DMA
IOR*	PLAT. DMA	SD	MEM. RES. (WR)
IOW*	PLAT. DMA	SD	ADD-ON SLAVE (RD)
IRQ	ANY RESOURCE	SMR/SMW*	PLAT. CIRCUIT
LA	PLAT. DMA	BCLK	PLAT. CIRCUIT
MASTER*	PLAT. CIRCUIT	TC	PLAT.. DMA (1)
MEMCS16*	PLAT. CIRCUIT	SRDY*	PLAT. CIRCUIT

TABLE 4-6: ADD-ON SLAVE CARD AS MEMORY RESOURCE DATA DESTINATION FOR TRANSFER CYCLES

LINE	DRIVEN BY	LINE	DRIVEN BY
AEN	PLAT. CIRCUIT	MEMR*	PLAT. DMA
BALE	PLAT. CIRCUIT	MEMW*	PLAT. DMA
DACK*	PLAT. DMA	OSC	PLAT. CIRCUIT
DRQ	I/O RESOURCE	REF*	PLAT. REF. CONT.
IOCHCK*	ALL RESOURCES	RESET	PLAT. CIRCUIT
IOCHRDY	ADD-ON SLAVE	SA	PLAT. DMA
IOCS16*	PLAT. CIRCUIT	SBHE*	PLAT. DMA
IOR*	PLAT. DMA	SD	I/O RES. (WR)
IOW*	PLAT. DMA	SD	ADD-ON SLAVE (RD)
IRQ	ANY RESOURCE	SMR/SMW*	PLAT. CIRCUIT
LA	PLAT. DMA	BCLK	PLAT. CIRCUIT
MASTER*	PLAT. CIRCUIT	TC	PLAT. DMA (1)
MEMCS16*	PLAT. CIRCUIT	SRDY*	PLAT. CIRCUIT

TABLE 4-7: ADD-ON SLAVE CARD AS MEMORY RESOURCE DATA SOURCE FOR TRANSFER CYCLES

NOTES:

(1): On an E-ISA platform, the TC signal line can also be driven by the add-on I/O slave card.

TRANSFER CYCLE DATA SIZE MISMATCH

The I/O resource, whether platform or add-on slave card, that uses one of the 8 bit DMA channels must be an 8 bit resource. The data will be transferred between the add-on slave I/O card and the memory resource only on the SD<0-7> signal lines. The memory resource, whether on the platform or on an add-on slave card, can be either 8 or 16 bits in size. The data will be mapped between the SD<0-7> and SD<8-15> signal lines by the platform byte swapping circuitry if the MEMCS16* signal line is activated by the memory resource and if the SA0 signal line equals logic 1 to indicate an odd address. Otherwise, the data will be only available on the SD<0-7> signal lines. In either case, the memory resource can directly map the byte into odd address locations. A transfer involving an even address memory byte requires both SA0 to be logic 0 and SBHE* to be inactive. The data will only be available on the SD<0-7> signal lines independent of the MEMCS16* signal line being activated. The data can be directly mapped into the even address location.

The I/O resource, whether platform or add-on slave card, that uses one of the 16 bit DMA channels must be a 16 bit resource. The data can only be transferred with a memory resource, whether on the platform or on an add-on slave card, that is also a 16 bit resource. No transfers are allowed between 16 bit I/O and 8 bit memory resources. The 16 bit transfer involving a memory word requires both the SA0 signal line to be logic 0 and the SBHE* signal line to be active. The data is available on the SD<0-15> signal lines, and can be directly mapped to the even memory address location.

> *Transfer cycles between 16 bit I/O resources and 16 bit memory resources do not require the MEMCS16* signal line to be activated. The transfer cycles between 8 bit I/O resources and 16 bit memory resources <u>do</u> require the MEMCS16* signal line to be activated to properly control the byte swapper circuitry.*

Table 4-8 outlines the interpretation by the platform byte swapping circuitry of specific signal lines during a transfer cycle.

I/O RESOURCE VALID DATA	MEM. RESOURCE VALID DATA	MEM. RESOURCE MEMCS16*	MEM. RESOURCE MAPPING ADDRESS	DMA SBHE*	DMA SA0	BYTE SWAP
8 BITS						
SD<0-7>	SD<0-7>	INACT.	EVEN BYTE	INACT.	0	NO
SD<0-7>	SD<0-7>	INACT.	ODD BYTE	X	1	NO
SD<0-7>	SD<0-7>	ACT.	EVEN BYTE	INACT.	0	NO
SD<0-7>	SD<8-15>	ACT.	ODD BYTE	ACT.	1	YES
16 BITS						
SD<0-15>	SD<0-15>	ACT	EVEN WORD	ACT.	0	NO

TABLE 4-8: BYTE SWAPPING FOR ISA AND E-ISA COMPATIBLE MEMORY AND I/O RESOURCES FOR TRANSFER CYCLES

See Chapter 7 for more detailed information.

> *The original IBM AT platform only uses the MEMCS16* signal line as an input. Even though the platform memory resource may be 16 bits, it is not able to drive the MEMCS16* signal line active. Thus, DMA controllers and the byte swapping circuitry do not monitor the MEMCS16* signal line on IBM AT platforms, and simply assume that the memory resource is 16 bits.*
>
> *Memory resources on ISA platforms also might not drive the MEMCS16* signal line. Assuming so may cause problems if an 8 bit add-on memory slave card is the memory resource in the transfer cycle. On an E-ISA platform, the MEMCS16* signal line is properly driven; consequently, the use of an 8 bit add-on memory slave card does not cause problems.*

ADD-ON BUS OWNER CARD

An add-on card can be designed as a bus owner. To become bus owner, it must first execute the obtain portion of the arbitration cycle. After it no longer has any immediate need for bus ownership, it can execute the release portion of the arbitration cycle. See Chapter 9 for more detailed information.

An add-on bus owner card has complete access to the entire memory and I/O address space. However, the platform architecture and software executing on the platform CPU assume that only the platform CPU has access to certain specific memory and I/O resources. For example, an add-on bus owner card can access the math co-processor memory registers; it can also access the I/O registers of the interrupt, DMA, and keyboard controllers, the timer, DMA page, real time clock, the NMI chip select circuitry, and Port B. Due to software compatibility issues, however, only the platform CPU should access the foregoing resources. These and other hardware limitations must be accounted for when writing software for add-on bus owner cards.

Of particular note is the fact that ISA and E-ISA bus ownership is awarded via the platform's DMA controller.

Throughout the add-on card bus ownership, the platform CPU and other platform resources assume that DMA transfer cycles are being executed. The activation of the MASTER* line by the add-on bus owner card immediately deactivates the AEN line, so I/O resources will respond to access cycles.

Access cycles are the only bus cycles that can be executed by the add-on bus owner. Add-on bus owner cards can be designed to execute access cycles in the same manner as the platform CPU. Resources on the bus are unable to determine if the platform CPU or an add-on bus owner card is executing the cycle unless the MASTER* line is monitored, and the add-on bus owner card cannot be assured that this line is being monitored. The single major difference between a platform CPU and an add-on bus owner card access cycle is the operation of BALE and the LA<17-23> lines. Since BALE is driven to logic 1 by platform circuitry, LA<17-23> must be driven valid by the add-on bus owner card for the entire access cycle, with the same timing as the SA<0-19> signal lines.

The original IBM AT platform architecture was designed with the MEMCS16, SRDY*, and IOCHRDY signal lines as inputs, and all were attached to +5 volts by pull-up resistors. Any access to the platform memory resources by an add-on bus owner card should only be executed as an 8 bit default memory cycle, but the IBM AT platform only allows a 16 bit access to its memory from the bus. Consequently, add-on bus owner cards are designed to only execute 16 bit memory access standard cycles and do not monitor the SRDY*, MEMCS16* and IOCHRDY signal lines. Similarly, the IOCS16* line is also an input attached to +5 volts by a pull-up resistor. As previously described, all of the I/O resources on the original IBM PC AT platform are not accessible by an add-on bus owner card; thus the only I/O resources available on the original IBM platforms to an add-on bus owner card are those on other add-on slave cards which support the IOCS16* signal line.*

For the above reasons, ISA compatible add-on bus owner cards only execute 16 bit standard access cycles to a 16 bit memory resource. These accesses ignore MEMCS16, SRDY*, and IOCHRDY signal lines, and execute only standard 16 bit cycles.*

The E-ISA compatible platform resources support the MEMCS16, SRDY*, and IOCHRDY signal lines. Consequently, E-ISA compatible add-on cards monitor these lines and execute 8 and 16 bit no wait state, standard, and ready access cycles accordingly. A jumper or configuration space option is needed to force 16 bit accesses independent of the MEMCS16* and IOCS16* signal lines to maintain compatibility with the IBM AT platform and other ISA platforms.*

E-ISA compatible add-on bus owner cards also monitor the IOCHRDY signal line, and will implement the cycles accordingly. This will insure optimum performance for all add-on slave cards on all platforms. The IBM AT platform and other "faithful" ISA clones which do not drive the IOCHRDY signal line will have no compatibility problem. All of the cycles will be executed as standard cycles.

Tables 4-9 and 4-10 outline which signal lines are driven by which resources during the two types of add-on bus owner card cycles. In the following tables, "PLAT. CIRCUIT" in some cases may simply be a pull-up resistor. "PLAT." is the abbreviation for "platform".

LINE	DRIVEN BY	LINE	DRIVEN BY
AEN	PLAT. CIRCUIT	MEMR*	ADD-ON BUS OWNER
BALE	PLAT. CIRCUIT	MEMW*	ADD-ON BUS OWNER
DACK*	PLAT. DMA	OSC	PLAT. CIRCUIT
DRQ	I/O RESOURCE	REF*	ADD-ON BUS OWNER
IOCHCK*	ALL RESOURCES	RESET	PLAT. CIRCUIT
IOCHRDY	MEMORY RESOURCE	SA	ADD-ON BUS OWNER
IOCS16*	PLAT. CIRCUIT	SBHE*	ADD-ON BUS OWNER
IOR*	PLAT. CIRCUIT	SD	ADD-ON BUS OWN.(WR)
IOW*	PLAT. CIRCUIT	SD	MEM. RESOURCE (RD)
IRQ	ANY RESOURCE	SMR/SMW*	PLAT. CIRCUIT
LA	ADD-ON BUS OWNER	BCLK	PLAT. CIRCUIT
MASTER*	ADD-ON BUS OWNER	TC	PLAT. DMA
MEMCS16*	MEMORY RESOURCE	SRDY*	PLAT. CIRCUIT

TABLE 4-9: ADD-ON BUS OWNER CARD DURING MEMORY ACCESS CYCLES

LINE	DRIVEN BY	LINE	DRIVEN BY
AEN	PLAT. CIRCUIT	MEMR*	PLAT. CIRCUIT
BALE	PLAT. CIRCUIT	MEMW*	PLAT. CIRCUIT
DACK*	PLAT. DMA	OSC	PLAT. CIRCUIT
DRQ	I/O RESOURCE	REF*	ADD-ON BUS OWNER
IOCHCK*	ALL RESOURCES	RESET	PLAT. CIRCUIT
IOCHRDY	I/O RESOURCE	SA	ADD-ON BUS OWNER
IOCS16*	I/O RESOURCE	SBHE*	ADD-ON BUS OWNER
IOR*	ADD-ON BUS OWNER	SD	ADD-ON BUS OWN.(WR)
IOW*	ADD-ON BUS OWNER	SD	MEM. RESOURCE (RD)
IRQ	ANY RESOURCE	SMR/SMW*	PLAT. CIRCUIT
LA	PLAT. CIRCUIT	BCLK	PLAT. CIRCUIT
MASTER*	ADD-ON BUS OWNER	TC	PLAT. DMA
MEMCS16*	PLAT. CIRCUIT	SRDY*	PLAT. CIRCUIT

TABLE 4-10: ADD-ON BUS OWNER CARD DURING I/O ACCESS CYCLES

Table 4-11 outlines the interpretation by the platform byte swapping circuitry of specific signal lines during an access cycle. The basic concepts are the same as those outlined in the ACCESS CYCLE DATA SIZE MISMATCH section of this chapter.

ADD-ON BUS OWNER			MEMORY OR I/O RESOURCE		BYTE SWAP	ON BUS	
VALID DATA	SBHE*	SA0	SIZE	XCS16*		VALID DATA	MAPPING ADDRESS
8 BITS							
SD<0-7>	INACT.	0	8	INACT.	NO	SD<0-7>	EVEN BYTE
SD<8-15>	ACT.	1	8	INACT.	YES	SD<0-7>	ODD BYTE
SD<0-7>	INACT.	0	16	ACT.	NO	SD<0-7>	EVEN BYTE
SD<8-15>	ACT.	1	16	ACT.	NO	SD<8-15>	ODD BYTE
16 BITS							
SD<0-15>	ACT.	0	8	INACT.	NO	SD<0-7>	ODD BYTE
SD<0-15>	ACT.	0	16	ACT.	NO	SD<0-15>	EVEN WORD

TABLE 4-11: BYTE SWAPPING FOR ADD-ON BUS OWNER CARD

> *The condition of SBHE* inactive and SA0 logical 1 is not allowed.*

Because bus ownership appears to the platform as a DMA cycle, no refresh cycles will be requested by the platform. The platform does not request refreshes because the design of the DMA controller, when programmed in the cascade mode, does not allow the DMA transfer to be interrupted. The DMA channels that are allowed to support add-on bus owner cards must be programmed in the cascade mode. It is the responsibility of the add-on bus owner card to request a refresh cycle every 15.6 microseconds in order to prevent any loss of memory. Some platforms are designed to automatically execute refresh cycles to platform DRAM independent of the bus. However, for DRAM resources on the bus, the refresh requests should be made by the add-on bus owner card. See Chapter 8 for more detailed information on the refresh bus cycles.

Table 4-12 outlines which signal lines are driven by which resources during a refresh cycle when an add-on bus owner card owns the bus. In the following table, "PLAT. CIRCUIT" in some cases may simply be a pull-up resistor. "PLAT." is the abbreviation for "platform".

LINE	DRIVEN BY	LINE	DRIVEN BY
AEN	PLAT. CIRCUIT	MEMR*	PLAT. REF. CONT.
BALE	PLAT. CIRCUIT	MEMW*	PLAT. CIRCUIT
DACK*	PLAT. DMA	OSC	PLAT. CIRCUIT
DRQ	I/O RESOURCE	REF*	ADD-ON BUS OWNER
IOCHCK*	ALL RESOURCES	RESET	PLAT. CIRCUIT
IOCHRDY	MEM. RESOURCES	SA	PLAT. REFRESH
IOCS16*	PLAT. CIRCUIT	SBHE*	PLAT. CIRCUIT
IOR*	PLAT. CIRCUIT	SD	TRISTATE
IOW*	PLAT. CIRCUIT	SD	TRISTATE
IRQ	ANY RESOURCE	SMR/SMW*	PLAT. CIRCUIT
LA	PLAT. CIRCUIT	BCLK	PLAT. CIRCUIT
MASTER*	ADD-ON BUS OWNER	TC	PLAT. DMA
MEMCS16*	PLAT. CIRCUIT	SRDY*	PLAT. CIRCUIT

TABLE 4-12: ADD-ON BUS OWNER CARD DURING REFRESH

REFRESH CYCLES

A refresh cycle can be executed by the refresh controller on the platform. The refresh cycle may have been requested by either the platform timer or by an add-on card that is bus owner. The add-on slave cards will respond to a refresh cycle in the same fashion independent of how the refresh cycle was requested. Table 4-13 outlines which signal lines are driven by which resource during a refresh cycle. In the following table, "PLAT. CIRCUIT" in some cases may simply be a pull-up resistor. "PLAT." is the abbreviation of "platform".

LINE	DRIVEN BY	LINE	DRIVEN BY
AEN	PLAT. CIRCUIT	MEMR*	PLAT. REF. CONT.
BALE	PLAT. CIRCUIT	MEMW*	PLAT. CIRCUIT
DACK*	PLAT. DMA	OSC	PLAT. CIRCUIT
DRQ	I/O RESOURCE	REF*	PLAT. REFRESH
IOCHCK*	ALL RESOURCES	RESET	PLAT. CIRCUIT
IOCHRDY	MEM. RESOURCES	SA	PLAT. REFRESH
IOCS16*	PLAT. CIRCUIT	SBHE*	PLAT. CIRCUIT
IOR*	PLAT. CIRCUIT	SD	TRISTATE
IOW*	PLAT. CIRCUIT	SD	TRISTATE
IRQ	ANY RESOURCE	SMR/SMW*	PLAT. CIRCUIT
LA	PLAT. CIRCUIT	BCLK	PLAT. CIRCUIT
MASTER*	PLAT. CIRCUIT	TC	PLAT. DMA
MEMCS16*	PLAT. CIRCUIT	SRDY*	PLAT. CIRCUIT

TABLE 4-13: ADD-ON SLAVE CARD DURING REFRESH

CHAPTER FIVE

SIGNAL LINE DEFINITION

INTRODUCTION

This chapter describes in detail all of the signal lines in an ISA or E-ISA compatible bus. The signals can be grouped into seven distinct categories: address, data, cycle control, bus control, interrupt, DMA, and power.

The implementation of the different bus signal lines vary for both the type of cycle and for the current bus owner. In some cases, it is platform circuitry that drives a particular bus line and not the current bus owner. The possible bus owners are the platform CPU, the DMA and refresh controllers, and add-on bus owner cards. Collectively, the non-platform CPU bus owners are the DMA and refresh controllers, and add-on bus owner cards.

ADDRESS SIGNALS

In addition to the actual address lines, other lines are used that aid in delivering the correct address-related information to all resources on the bus. These other signal lines are BALE, AEN, and SBHE*.

BALE (8 and 8/16 bit slots)

Bus address latch enable (BALE) is driven by the platform CPU to indicate when the ADDRESS, AEN, and SBHE* signal lines are valid. For non-platform CPU bus owners, this line is driven by platform circuitry to a constant logical one for the entire cycle.

> *For SAFE COMPATIBILITY, the input circuitry of an add-on slave memory card to latch the LA<17-23> signal lines for either a platform CPU or an add-on bus owner card should be as shown in Figure 5-1.*

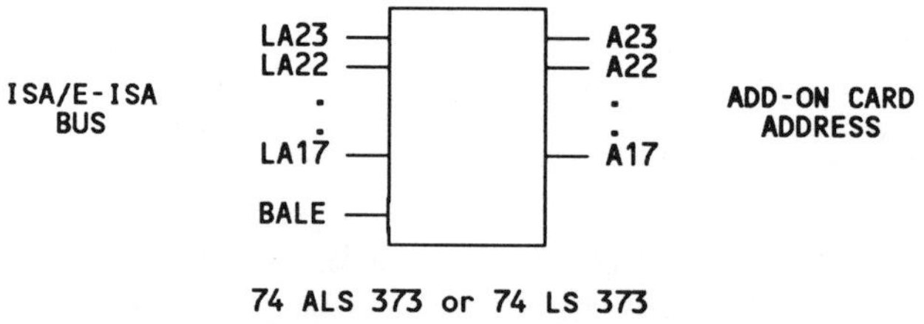

FIGURE 5-1: INPUT ADDRESS CIRCUIT

SA<0-19> (8 and 8/16 bit slots)

The SA<0-19> signal lines are driven by the current bus owner and define the lower 20 ADDRESS signal lines needed for the lower 1 megabyte of memory address space. All signal lines are driven for both memory and DMA cycles. The SA<0-15> signal lines define the entire 64K address space required for the I/O address space and are the only ones needed for I/O access cycles. When the platform CPU is bus owner, it must drive the ADDRESS signal lines prior to activating the BALE signal line. For non-platform CPU bus owners, the ADDRESS signal lines must be valid prior to and during the active period of the COMMAND signal lines. This requirement is because the platform circuitry drives the BALE signal line to a constant logical one for the entire cycle.

> *An add-on bus owner card may request a refresh cycle while it is the current bus owner. During the refresh cycle it must make the address lines inputs, because the refresh controller is driving the refresh address onto the bus.*

During refresh cycles, only the SA<0-7> signal lines are driven by ISA platforms. E-ISA platforms have expanded the definition of refresh to drive the SA<0-15> signal lines during the refresh cycle.

Any individual address lines that are not driven by the current bus owner will float to an undefined logical state.

LA<17-23> (8/16 bit slots)

The LA<17-23> signal lines are driven by the current bus owner and provide the additional address lines required for the 16 megabyte memory address space. When the platform CPU is bus owner, it only drives the ADDRESS signal lines for a small portion of the access cycle. The platform CPU relies on BALE to indicate to memory resources when these ADDRESS signal lines can be latched. For non-platform CPU bus owners, the ADDRESS signal lines must be valid prior to and after the active period of the COMMAND signal lines. This requirement is because the platform circuitry drives the BALE signal line to a constant logical one for the entire cycle.

During a refresh cycle, the LA<17-23> signal lines are undefined. All resources should tristate these lines during a refresh cycle.

Any individual address lines that are not driven by the current bus owner will float to an undefined logical state.

SBHE* (8/16 bit slots)

The system byte high enable (SBHE*) signal line is driven by the current bus owner to indicate that valid data resides on the SD<8-15> signal lines. This line, in conjunction with the SA0 signal line, allows the current bus owner to indicate to platform and bus resources an 8 versus 16 bit access or transfer cycle.

When the platform CPU is bus owner, it must drive the ADDRESS signal lines prior to activating the BALE signal line. For non-platform CPU bus owners, the SBHE* signal line must be valid prior to and during the active period of the COMMAND signal lines. This requirement is because platform circuitry drives the BALE signal line to a constant logical one for the entire cycle.

> *The SBHE* signal line is undefined for refresh cycles because no data is actually read.*

AEN (8 and 8/16 bit slots)

The address enable (AEN) signal line is driven by platform circuitry as indication to I/O resources not to respond to the ADDRESS and I/O COMMAND signal lines. This signal line has no effect on memory resources.

On ISA platforms, this signal line is inactive when the primary CPU or an add-on card is bus owner. This signal line is active only when the DMA controller is the bus owner. Only the DMA controller executes transfer cycles with the memory and the I/O COMMAND signal lines active simultaneously. The AEN signal line is the method by which I/O resources are informed that a DMA transfer cycle is occurring and that only the I/O resource with an active DACK* signal line can respond to the I/O signal lines.

On E-ISA platforms, the interpretation of this line for DMA transfer cycles is the same as for the ISA platform, except for one difference: an E-ISA platform supports slot-specific I/O addressing and configuration space. Consequently, the platform circuitry individually drives the AENx signal line to each slot. (The appended x indicates the slot number.) During a DMA transfer cycle, all of the AENx signal lines are active. An I/O access cycle executed by the platform CPU or add-on bus owner card decodes the I/O address and drives the appropriate AENx signal lines. If the access is to the general I/O address block, then all of the AENx signal lines are inactive. If the access is to a specific I/O address block, then only the AENx signal line associated with the slot is inactive.

See Chapter 2 for more detailed address space information, and Chapter 12 for more configuration space and slot specific I/O information.

For an ISA platform I/O access cycle, when the platform CPU is bus owner, the AEN signal line must be inactive prior to the BALE signal line becoming active, and must remain inactive during the active period of the COMMAND lines. When an add-on card is bus owner, the AEN signal line must be inactive prior to and during the active period of the COMMAND signal lines. This requirement is because the platform circuitry drives the BALE signal line to a constant logical one for the entire cycle.

For an E-ISA platform access cycle, the logic levels of all the AENx signal lines are the same as the just-described AEN signal line for the ISA platform. When the access is to a specific I/O address, the associated AENx signal line is inactive and all others are active.

For DMA transfer on either an ISA or E-ISA platform, the AEN signal line must be active prior to and during the active period of the COMMAND signal lines. When the platform CPU is bus owner, the appropriate AENx signal line must be valid prior to an active BALE and during the active period of the COMMAND signal lines.

> *When a refresh cycle is executed, the AEN signal line is active unless an add-on bus owner card requested the refresh cycle. There is no useful interpretation of this signal line during a refresh cycle and it should be ignored.*

DATA SIGNALS

SD<0-7> (8 and 8/16 bit slots)
SD<8-15> (8/16 bit slots)

The SD15 signal line represents the most significant bit and SD0 the least significant bit. The platform byte swapper circuitry insures that the SD<8-15> signal lines are valid whenever the SBHE* signal line is active. The SD<0-7> signal lines are valid during a memory (or I/O) access cycle, except when the SBHE* signal line is active and the SA0 signal line is logical 1 and the MEMCS16* (or IOCS16*) signal line is active. For DMA transfer cycles, the SD<0-7> signal lines are always valid.

> *The SD<0-15> signal lines are undefined for a refresh cycle. All resources should tristate these signal lines during a refresh cycle.*

CYCLE CONTROL SIGNALS

These signals control the address space (memory or I/O), data direction (read or write), and the length of the cycle.

MEMR* (8/16 bit slots)
SMEMR* (8 and 8/16 bit slots)

The memory read (MEMR*) signal line is driven by the current bus owner to request a memory resource to drive data onto the bus during the cycle. The system memory read (SMEMR*) signal line does the same thing, except it is active only within the first 1 megabyte of the address space. The SMEMR* signal line is active when the MEMR* signal line is active and the LA<20-23> signal lines indicate the first 1 megabyte of address space. It is driven by the platform circuitry monitoring the MEMR* and LA<20-23> signal lines; consequently, its timing lags that of the MEMR* signal line.

> *During a refresh cycle, the MEMR* signal line is driven by the refresh controller.*

MEMW* (8/16 bit slot)
SMEMW* (8 and 8/16 bit slots)

The memory write (MEMW*) and the system memory write (SMEMW*) signal lines request the memory resource to accept data from the data lines. Otherwise, they are defined in the same fashion as the MEMR* and SMEMR* signal lines.

> *During a refresh cycle, these signal lines are not active. Pull-up resistors on the platform keep these signal lines from floating active.*

IOR* (8 and 8/16 bit slots)

The I/O read (IOR*) signal line is driven by the current bus owner to request an I/O resource to drive data onto the data bus during the cycle. The I/O resource is responding to a valid address for an access cycle, or to a valid DACK* signal line for a transfer cycle.

> *During a refresh cycle, this signal line is not active. Pull-up resistors on the platform keep this signal line from floating active.*

IOW* (8 and 8/16 bit slots)

The I/O write (IOW*) signal line is driven by the current bus owner to request an I/O resource to accept data from the data lines. The I/O resource responds to a valid address for an access cycle, or to a valid DACK* signal line for a transfer cycle.

> *During a refresh cycle, this signal line is not active. Pull-up resistors on the platform keep this signal line from floating active.*

MEMCS16* (8/16 bit slots)

The memory chip select 16 (MEMCS16*) signal line is driven by the memory resource if it can support a 16 bit access cycle. An active MEMCS16* signal line not only indicates support of 16 bit data; it also allows the bus owner to execute shorter cycles. The MEMCS16* signal results from a straight decode of the LA<17-23> signal lines without any qualification of the COMMAND signal lines. Consequently, it will be driven active also during I/O access cycles. The bus owner must ignore this line during I/O access cycles.

On an ISA or E-ISA platform, the MEMCS16* signal line is driven active within a specific time of a valid address to execute a 16 bit cycle. This event occurs asynchronously to the BCLK signal line.

> *The MEMCS16* signal line is ignored by the DMA and refresh controller for transfer and refresh cycles, respectively.*

> *The original IBM AT and ISA platforms do not drive the MEMCS16* signal line. Therefore, an access to the platform memory by an add-on bus owner card should ignore the MEMCS16* signal line and execute the access as a 16 bit cycle. E-ISA platforms properly drive the MEMCS16* signal line when platform memory is accessed. See the ADD-ON BUS MASTER CARD section of Chapter 4 for more detailed information.*

> *The fact that MEMCS16* is a straight decode of the LA<17-23> signal lines means that the entire 128K memory address block is either all 8 bits or all 16 bits. Problems arise if only a portion of the block is associated with a 16 bit resource and another portion is an 8 bit resource. When the platform CPU is the bus owner the SA<0-16> signal lines do not become valid soon enough to allow for different bit sizes within the 128K block.*

IOCS16* (8/16 bit slots)

The I/O chip select 16 (IOCS16*) signal line is driven by an I/O resource if it can support a 16 bit access cycle. An active IOCS16* signal line not only indicates a support of 16 bit data; it also allows the bus owner to execute shorter default cycles. IOCS16* is a straight decode of the SA<0-15> signal lines without any qualification of the COMMAND signal lines. Consequently, the IOCS16* signal line may be driven active during memory access cycles. The current bus owner must ignore this line during memory access cycles.

On an ISA or E-ISA platform, the IOCS16* signal line must be driven inactive within a specific time of a valid address to execute a 16 bit cycle. This event occurs asynchronously to the BCLK signal line.

> *The IOCS16* signal line is ignored by the DMA and refresh controller for transfer and refresh cycles, respectively.*

> *An E-ISA compatible add-on card must not activate the IOCS16* signal line if the AEN signal line of the associated slot is active.*

> *The original IBM AT baseboard does not drive the IOCS16* signal line because I/O platform access is not allowed. Consequently, ISA compatible add-on bus owner cards will execute 16 bit access cycles without monitoring the IOCS16* signal line. E-ISA platforms properly drive the IOCS16* signal line when platform I/O is accessed. See the ADD-ON BUS MASTER CARD section of Chapter 4 for more detailed information.*

IOCHRDY (8 and 8/16 bit slots)

The I/O channel ready (IOCHRDY) signal line allows resources to indicate to the bus owner that additional cycle time is required. If the IOCHRDY signal line is not driven inactive, the cycle is either a no wait state or a standard cycle. If the IOCHRDY signal line is driven inactive, the cycle length is extended until the line is driven active again. An inactive IOCHRDY signal line supersedes an active SRDY* signal.

On an ISA and E-ISA platform, the IOCHRDY signal line has to be driven inactive within a specific time of the COMMAND signal lines becoming active. When the

IOCHRDY signal line is active, the cycle ends within a specific amount of time. Both of these events occur asynchronously to the BCLK signal line. On an E-ISA platform, when the platform CPU or DMA controller is bus owner, additional information relates the IOCHRDY signal line sampling with the BCLK signal line. See Chapters 6 and 7 for more detailed information.

When the bus owner is either the platform CPU, an add-on bus owner card, or the refresh controller, the increase in cycle length is measured as "wait states". The resolution of the wait states are BCLK periods. When the bus owner is the DMA controller, the resolution of the wait states is two BCLK periods for an ISA compatible cycle. The DMA controller on an E-ISA platform allows each DMA channel to be programmed for different DMA transfer cycles. Each of these cycles has a different wait state resolution. See the DMA TRANSFER CYCLE section in Chapter 6 for more detailed information.

For a DMA transfer cycle, only the memory resource can inactivate the IOCHRDY signal line. The I/O resource must be able to accept the data within the length of a default cycle.

> *The IOCHRDY signal line should not remain inactive during access or transfer cycles longer than 15.6 microseconds. Otherwise, a bus refresh cycle may not occur at the appropriate time and DRAM data will be lost. The IBM Technical Reference Manual specifies 2.5 microseconds for unknown reasons.*

IOCHRDY SIGNAL LINE ENHANCEMENT

As previously described, the IOCHRDY signal line must be inactive within a "specific" time after the COMMAND signal line becomes active for a ready cycle. The ready cycle is completed at some time 125 nanoseconds (120 nanoseconds for E-ISA) after the IOCHRDY signal line has been driven active. This timing of the IOCHRDY signal line is fairly "imprecise" and tends to lower performance.

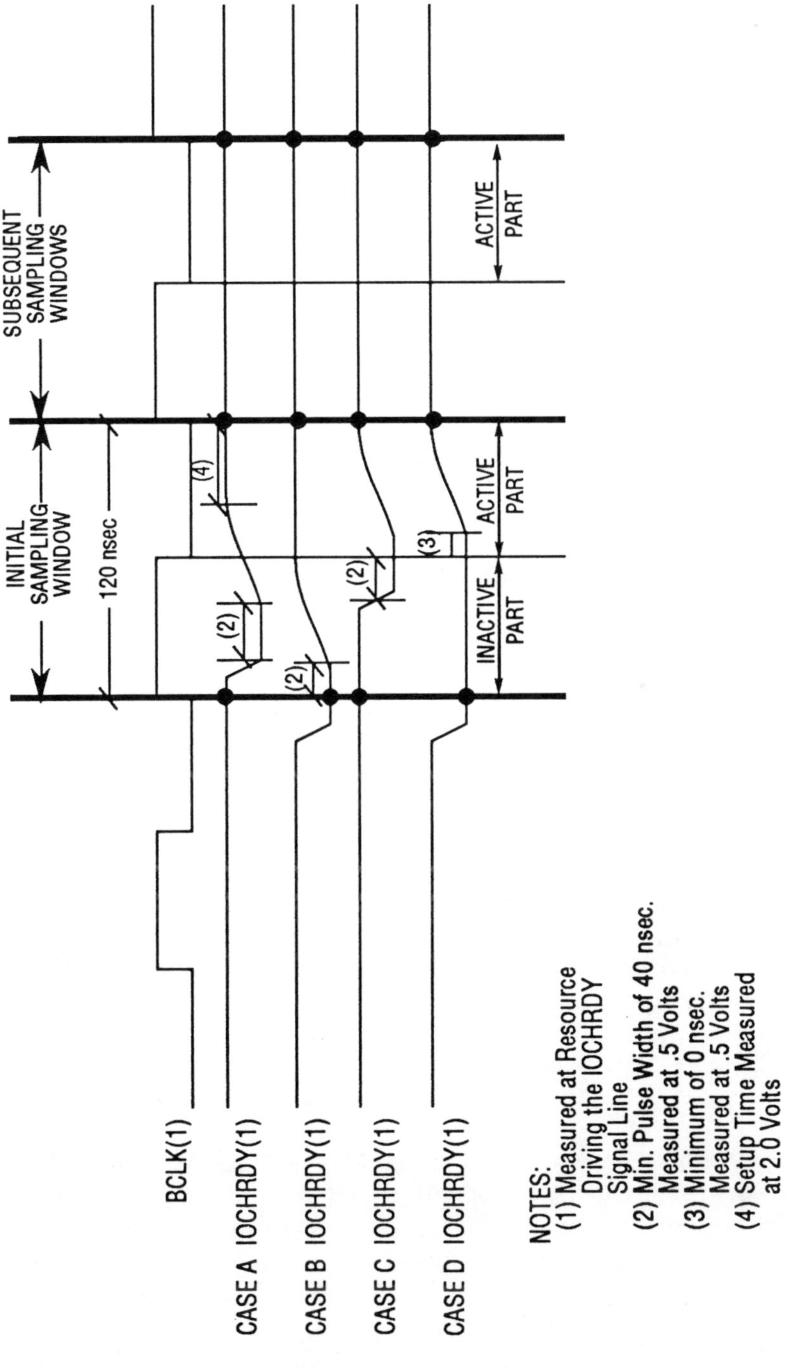

FIGURE 5-2: E-ISA IOCHRDY REFERENCE TO THE BCLK SIGNAL LINE

In order to enhance the performance, the EISA bus specification has defined a SAMPLING WINDOW that relates the IOCHRDY and BCLK signal lines. Figure 5-2 outlines this SAMPLING WINDOW as viewed by the access or transfer cycle resource. The SAMPLING WINDOW can be viewed in two parts: inactive and active.

The inactive part of the INITIAL SAMPLING WINDOW is between the rising and falling edges of the BCLK signal line. An IOCHRDY signal line inactive pulse of at least 40 nanoseconds within the inactive part of the window will cause a ready cycle to be executed. For CASE C, the minimum setup time of 40 nanoseconds to the falling edge of the BCLK signal line requires the receiving platform CPU or add-on bus owner card to support an actual setup time of 10 nanoseconds. The 30 nanosecond loss reflects allowances for bus settling and clock tolerances.

In the active part of the INITIAL SAMPLING WINDOW, the IOCHRDY signal line is sampled at the rising clock edge at the end of the INITIAL SAMPLING WINDOW. If the IOCHRDY signal line is sampled active, the ready cycle will be completed at the end of the next BCLK period. The EISA bus specification states that a setup time of 34 nanoseconds is required. However, the IOCHRDY signal line is specified as an open collector device with a 1K pull-up resistor. Assuming a 10 nanosecond setup time at the platform CPU or add-on bus owner card, an 11 nanosecond BCLK signal line settling time, and a 78 nanosecond rise time, the setup time relative to the open collector driver is 99 nanoseconds. With a minimum INITIAL SAMPLING WINDOW of 120 nanoseconds and 40 nanosecond minimum pulse width, none of the CASES will result in a guaranteed one wait state ready cycle. The situation becomes worse if the value of 20 pf per slot or 16 fully loaded slots as allowed by the EISA bus specification is considered.

To guarantee a one wait state ready cycle, the IOCHRDY signal line must be pulsed high, or a 300 ohm pull-up resistor must be used on the bus. However, a fully loaded 16 slot backplane as allowed by the EISA bus specification is still a problem.

If the IOCHRDY signal line is not sampled active at the end of the INITIAL SAMPLING WINDOW, then SUBSEQUENT SAMPLING WINDOWS occur. These SUBSEQUENT SAMPLING WINDOWS only have an active part. The IOCHRDY signal line is sampled at the rising clock edge at the end of each sampling window. If the IOCHRDY signal line is sampled active, the ready cycle will be completed at the end of the next BCLK period. The problems outlined for the active part of the INITIAL SAMPLING WINDOW also apply to the active part of the SUBSEQUENT SAMPLING WINDOWS.

An add-on slave card that implements the 40 nanosecond IOCHRDY signal line inactive pulse will not work with an ISA compatible platform CPU or DMA controller.

SRDY* (8 and 8/16 bit slots)

The synchronous ready (SRDY*) signal line is driven active by the accessed resource to indicate that an access cycle shorter than default can be executed. In an ISA system, only the platform CPU can execute a no wait state cycle. An inactive IOCHRDY signal line supersedes an active SRDY* signal forcing a ready access cycle instead of a no wait state cycle.

On ISA and E-ISA platforms, the SRDY* signal line is synchronized to the BCLK signal line. It must be driven active within a specific time of the COMMAND signal lines being activated during 16 bit no wait state memory access cycles.

E-ISA has expanded the definition of no wait state cycles to include all platform CPU and add-on bus owner card access cycles to 8 bit memory and I/O resources. Only the 16 bit I/O access cycle does not support no wait state cycles. Also, on the E-ISA platform, the SRDY* signal line for 8 bit accesses is sampled at two places. See Chapter 6 for more detailed information. (Other names which are used for SRDY are 0WS* and ENDXFR*.)

> *The SRDY* signal is ignored by the DMA and refresh controller during transfer and refresh cycles.*

The original IBM AT running the BCLK signal line at 8 Mhz cannot support no wait state cycles. If bus losses are assumed to be 0 nanoseconds, then it appears that the add-on card has 20 nanoseconds to drive the SRDY signal line active from an active memory COMMAND signal line. However, if 22 nanoseconds of bus losses are included (11 nanoseconds for COMMAND signal line active at the CPU to add-on slave and 11 nanoseconds for SRDY* to arrive at the CPU), then the add-on card has -2 nanoseconds to drive the SRDY* signal line active to meet the setup time to the falling edge of the BCLK signal line. The ability to run no wait state cycles at an 8 Mhz BCLK signal line clock rate on original ATs is due to "grace" of typical delay, typical settling, and typical setup times. The worst case collection of these times will cause a problem. Very few no wait state add-on cards have been built.*

The original IBM AT running the BCLK signal line at 6 Mhz can support no wait state cycles. If the BCLK signal line clock rate is 6 Mhz, then the add-on card has 19 nanoseconds (including 22 nanoseconds of bus losses) to drive the SRDY signal line active from an active memory COMMAND signal line. As outlined above, these 19 nanoseconds disappear at an 8 Mhz BCLK signal line clock rate.*

ISA and E-ISA platforms have clock rates of 8 or 8.33 Mhz, respectively. These platforms must have improved COMMAND signal lines delays and setups to falling BCLK signal line clock edge to support no wait state cycles. The EISA bus specification has established a maximum delay of 15 nanoseconds from an active COMMAND signal line to activating the SRDY signal line. Also, the EISA bus specification has established a minimum of 10 nanoseconds to the falling edge of the BCLK signal line relative to an active SRDY* signal line. (The IEEE P996 specification has less restrictive times. See the timing table in Chapter 6.)*

The timing table in Chapter 6 also specifies additional time relationships between the ADDRESS signal lines and the SRDY signal line. The adherence to these time relationships will insure that the SRDY* signal line active setup time to the BCLK signal line falling edge is met.*

BUS CONTROL SIGNALS

This group of signals provides general cycle support and bus control.

REFRESH* (8 and 8/16 bit slots)

The memory refresh (REFRESH*) signal line is driven by the refresh controller to indicate a refresh cycle. If an add-on bus owner card is the current bus owner, then it can request a refresh cycle from the refresh controller by activating the REFRESH* line. In order to prevent loss of DRAM data, the refresh cycle must be requested every 15.6 microseconds.

MASTER* (8/16 bit slots)

The MASTER* signal line is only driven active by an add-on bus owner card that wants to become the bus owner. It can only be driven after the add-on bus owner card has activated the DACK* signal line. The add-on bus owner remains bus owner until it drives the MASTER* signal line inactive.

During a refresh cycle requested by an add-on bus owner card, the MASTER signal line is active. Otherwise, during a refresh cycle, a platform pull-up resistor keeps the signal line inactive.*

IOCHCK* (8 and 8/16 bit slots)

The I/O channel check (IOCHCK*) signal line can be driven active by any resource. It is a general error condition that has no specific interpretation. It can occur due to parity errors, general bus failure, and other types of errors. This signal can only be serviced by the platform CPU.

The active state of the IOCHCK* signal line is latched into port B of the platform. A non-maskable interrupt (NMI) is issued to the platform CPU if bit 3 of port B (61H) is set to "1"; otherwise, the primary CPU software must poll the IOCHCK bit in port B.

When either the add-on bus owner, refresh controller, or DMA controller is the current bus owner, the activation of the IOCHCK* channel check signal will not be acted upon immediately by the platform CPU. The platform CPU must wait until it becomes the bus owner again prior to executing any error correction software.

RESET (8 and 8/16 bit slots)

The RESET signal line is driven active by platform circuitry. Any bus resource that senses an active reset line must immediately tristate all output drivers and enter the appropriate reset condition.

The RESET signal line can be activated at any time, by either the platform hardware or the primary CPU setting port B (61H) bit 0 to a "1". The minimum pulse width is 1 microsecond. See Chapter 6 for more detailed information.

> *In order for software to move between protected and real mode operation, the primary CPU must be reset. This reset does not affect the condition of the RESET signal line.*

BCLK (8 and 8/16 bit slots)

The system bus clock (BCLK) signal line is a clock driven by platform circuitry. On ISA platforms it has a 50% (+/- 5%) duty cycle of 6 to 8 Mhz (+/- 500 ppm). On E-ISA platform, the clock has a 50% (+/- 5%) duty cycle of 6 to 8.33 Mhz (+/- 500 ppm). It is not synchronized to any other bus signal lines except SRDY*. For E-ISA platforms this has been expanded for reference sampling of IOCHRDY and DRQ signal lines.

OSC (8 and 8/16 bit slots)

The oscillator (OSC) signal line is a clock driven by platform circuitry. It has a 45-55% duty cycle of 14.31818 Mhz (+/- 500 ppm). It is not synchronized to any other bus signal line.

This frequency is the color-burst frequency for television, and results in the availability of large numbers of inexpensive crystals. On the platform, this clock is divided by 12 and fed to the 8254 timer.

INTERRUPT SIGNALS

The INTERRUPT signal lines allow add-on cards to request interrupt service by the platform CPU. Only the platform CPU can be notified of the interrupt by the interrupt controller. Also, only the platform CPU can access the registers in the interrupt controller.

IRQ<3-7,9> (8 and 8/16 bit slots)
IRQ<10-12,14,15> (8/16 bit slots)

The INTERRUPT signal lines can be programmed for one type of operation on an ISA platform and two types of operation on an E-ISA platform. On the ISA platform, the INTERRUPT signal lines are considered active by a low to high transition (edge triggered); consequently, only one add-on card can be assigned to each INTERRUPT signal line. The signal line must remain active until the platform CPU has accessed the appropriate portion of the add-on card to indicate that the interrupt is being serviced.

On the E-ISA platform, the INTERRUPT signal lines can be programmed to be edge or level triggered. When programmed to be edge triggered, their operation is the same as on the ISA platform. When programmed to be level triggered, the INTERRUPT signal lines must be driven to an active low to be considered active. This approach allows several add-on cards to use the same INTERRUPT signal line if the line is driven by open collector drivers. The signal line must remain active until the platform CPU has accessed the appropriate portion of the add-on card to indicate that the interrupt is being serviced.

> *The IBM AT and ISA platforms have defined the INTERRUPT signal lines' operation as unshared edge triggered; consequently, add-on cards have been designed with totem pole outputs. If these add-on cards are installed in an E-ISA platform, they must be the only interrupt resources on the line.*
>
> *The open collector approach for a level triggered INTERRUPT signal line relies on a pull-up resistor to deactivate the interrupt request. The EISA bus specification has the largest resistor value; consequently, 500 nanoseconds are required for deactivation on the E-ISA platform.*

> *The INTERRUPT signal lines are being continuously monitored by the interrupt controller independent of the present cycle or bus owner.*

DMA SIGNALS

Each DMA channel has two signal lines: DMA request (DRQ) and DMA Acknowledge (DACK*). Also, all DMA channels share the terminal count (TC) signal line. The purpose of these signal lines is to allow data to be transferred between I/O and memory resources. The platform DMA controller drives all of the control lines to execute the cycle; consequently, no control circuitry is required on the memory or I/O resource.

DRQ <1-3> (8 and 8/16 bit slots)
DRQ <0,5-7> (8/16 bit slots)

The DRQ signal lines are driven active by I/O resources to request service by the platform DMA controller. They must remain active until the appropriate DACK* signal line is driven active.

The DRQ signal lines are also driven active when an add-on bus owner card wants to own the bus. The add-on bus owner drives the MASTER* signal line active and takes control of the bus once the appropriate DACK* signal line is driven active.

> *On the IBM AT and ISA platforms, the DRQ signal lines are defined to have totem pole type of outputs; thus, add-on cards require a jumper to attach the driver to the appropriate DRQ signal line. The EISA bus specification has defined the DRQ signal lines to have tristate drivers; consequently, so do E-ISA compatible add-on cards.*

As outlined in the SPECIAL DMA CONSIDERATIONS section of Chapter 7, four DMA modes are supported. Each mode must sample when the DRQ signal line is active or inactive. For a standard transfer cycle, the sample point is the rising edge of the BCLK signal line. The specific BCLK signal line period at which the sampling takes place varies for the COMPATIBLE, TYPE "A", AND TYPE "B" transfer cycles. (See Figures 5-3 to 5-5.) The EISA bus specification has established a 30 nanosecond output delay for the COMMAND signal line active from a rising edge of the BCLK signal line (#3). It has also specified a 15 nanosecond setup time for valid DRQ signal line to the rising edge of the BCLK signal line (#4). If a bus settling time of 11 nanoseconds is assumed, then 22 nanoseconds total are required for bus settling (COMMAND signal line from the DMA controller to the add-on slave card, plus the DRQ signal line from the add-on slave card to the DMA controller). The total of output, settling, and setup is 30 + 22 + 15 = 67 nanoseconds. The EISA bus specification assumes only a 60 nanosecond total time for output, settling, and setup.

Based on the above information, the delay for a valid DRQ signal line from an active COMMAND signal line is summarized in Table 5-1 for Figures 5-3 to 5-5.

TRANSFER CYCLE TYPE	# (4)	MAX. (NSEC) ISA (1)	E-ISA (2)	EISA (3)
COMPATIBLE	1	558	533	540
	2	308	293	300
TYPE "A"	1	NA	293	300
	2	NA	173	180
TYPE "B"	1	NA	173	180
	2	NA	53	60

TABLE 5-1: DRQ SIGNAL LINE DELAY FROM COMMAND SIGNAL LINE

NOTES:

(1) BCLK = 8.00 MHZ
(2) BCLK = 8.33 MHZ
(3) BCLK = 8.33 MHZ
(4) Refer to numbers in Figures 5-3 through 5-8.

When the transfer cycle is of the ready type, the sample point for the DRQ signal line changes by wait state increments. For example, if the cycle is a COMPATIBLE one wait state ready cycle, the sample point moves to the beginning of the eighth clock period of the BCLK signal line. (See Figures 7-8 and 7-9.) For a TYPE "A" or "B" one wait state ready cycle, it moves to the beginning of the fifth and fourth BCLK signal line periods, respectively. (See Figures 7-10 and 7-11.) Remember that the wait state increment for a COMPATIBLE cycle is two BCLK periods; for a TYPE "A" or "B" it is one BCLK period.

The DMA REQUEST (DRQ) signal lines must be continuously monitored by the DMA controller independent of the present cycle or bus owner.

> *The IEEE P996 specification identifies an active IOW* signal line to a valid DRQ signal line delay of 118 nanoseconds. This reflects the fact that the output delay from BCLK of an 8237A-5 is 190 nanoseconds (308 - 190 = 118). The IEEE P996 also uses 118 nanoseconds for the delay of active IOR* signal line to a valid DRQ signal line.*

DACK* <1-3> (8 and 8/16 bit slots)
DACK* <0,5-7> (8/16 bit slots)

The DACK* signal lines are driven active by the platform DMA controller to select which add-on card will either be part of the transfer cycle or will be granted bus ownership.

TC (8 and 8/16 bit slots)

The TC signal line on an ISA platform is driven by the platform DMA controller to indicate that all data has been transferred. On an E-ISA platform, this signal line is bi-directional. It can be programmed to be driven by the platform DMA controller in the same fashion as the ISA platform. It can also be programmed for the add-on card to drive this signal line to terminate any further data transfers.

When programmed as an input, the active period of the TC signal line must be valid relative to active COMMAND signal lines and inactive DACK* signal lines. (See Figures 5-6 to 5-8.) The EISA bus specification has established a 30 nanosecond output delay of the COMMAND signal line from the rising edge of the BCLK signal line (#3). It has also specified a 15 nanosecond setup time for a valid TC signal line to the rising edge of the BCLK signal line (#4). If 11 nanoseconds is assumed for bus settling time (each direction), then 22 nanoseconds total are required for bus settling (COMMAND signal line from the DMA controller to add-on slave card plus TC signal line from the add-on slave card to the DMA controller). The total of output, settling, and setup is 30 + 22 + 15 = 67 nanoseconds. The EISA bus specification assumes 40 nanoseconds for COMPATIBLE and TYPE "A", and 50 nanoseconds for TYPE "B".

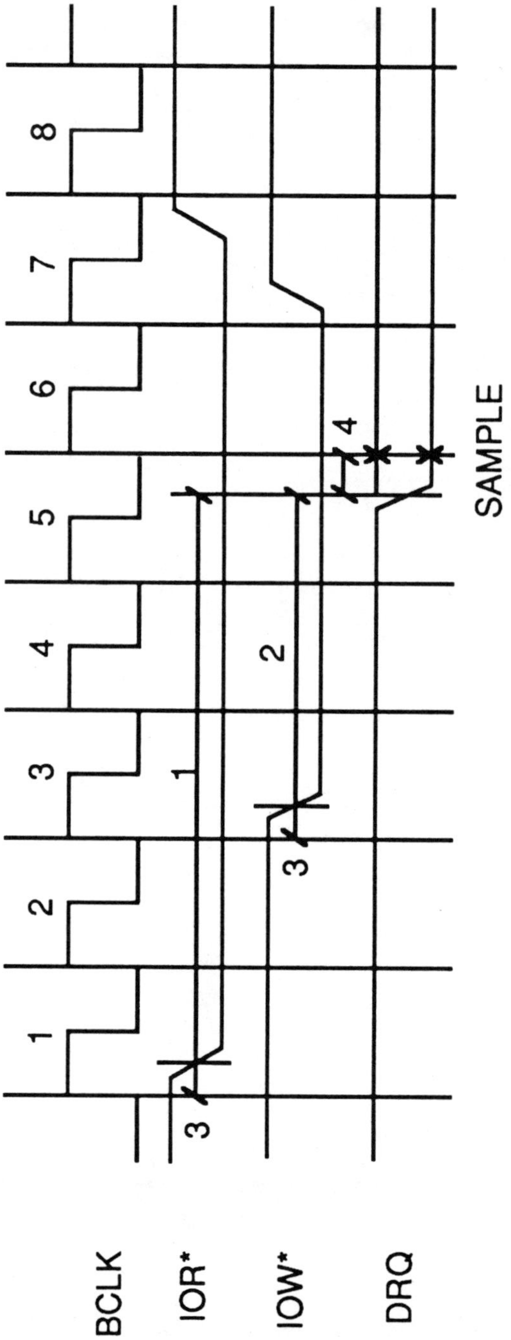

FIGURE 5-3: DRQ SIGNAL LINE TIMING
COMPATIBLE TRANSFER CYCLE (STANDARD CYCLE)

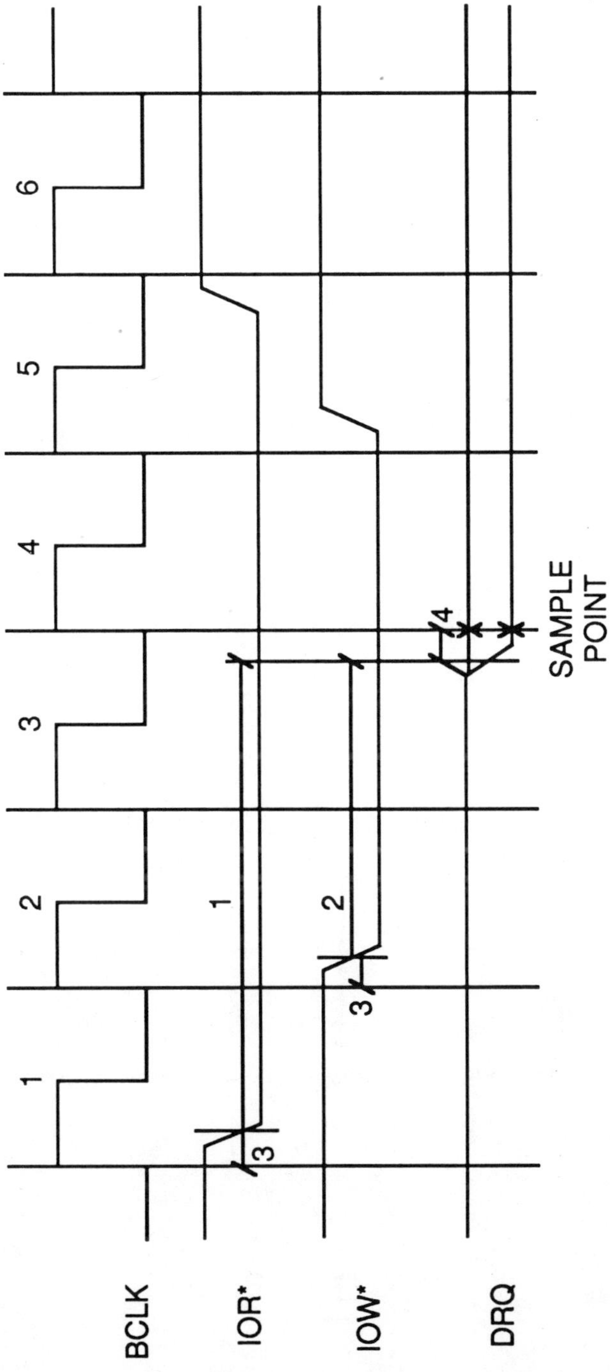

FIGURE 5-4: DRQ SIGNAL LINE TIMING
TYPE "A" TRANSFER CYCLE (STANDARD CYCLE)

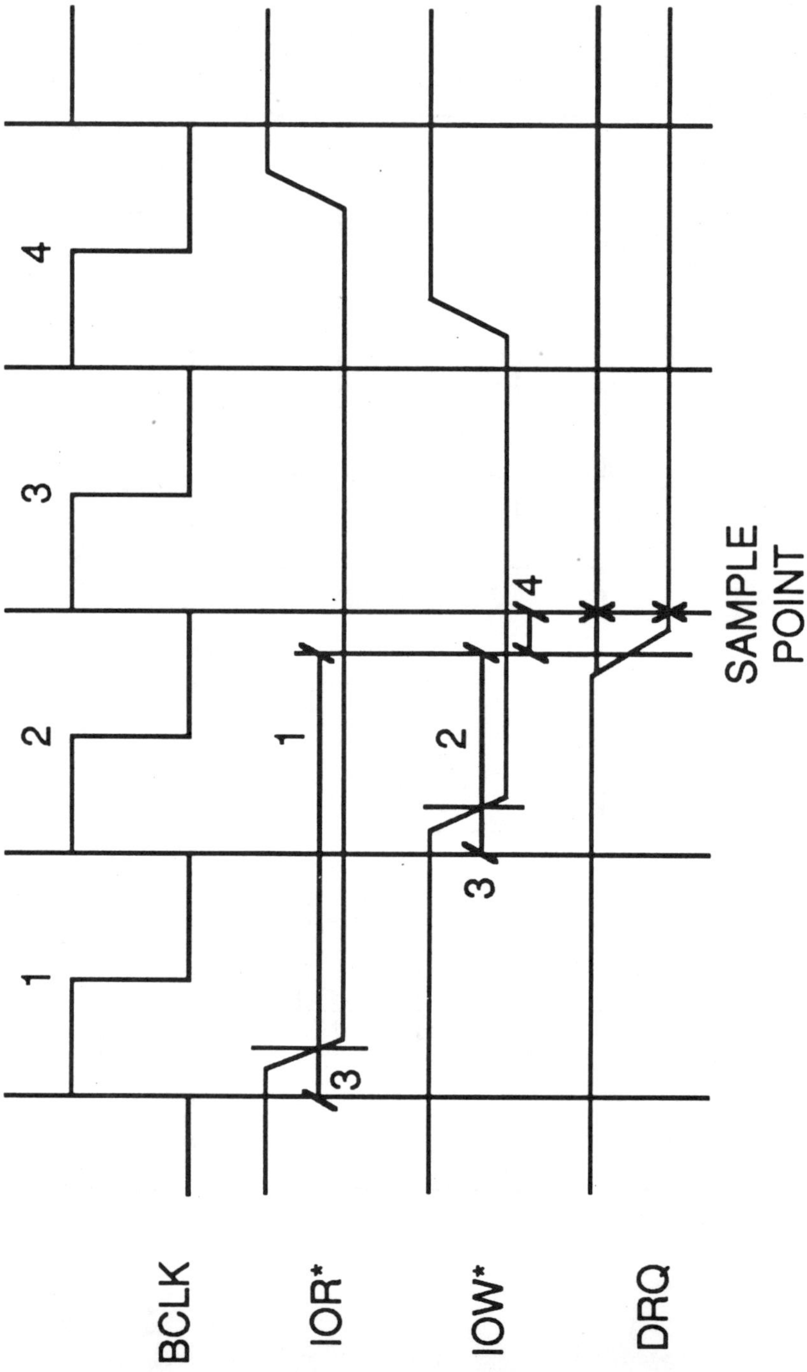

FIGURE 5-5: DRQ SIGNAL LINE TIMING
TYPE "B" TRANSFER CYCLE (STANDARD CYCLE)

The minimum hold time (5) of the TC signal line from an inactive DACK signal line is 40 nanoseconds (measured at the add-on card).

Given the above information, the delay for a valid TC signal from the active COMMAND signal line is summarized in Table 5-1 for Figures 5-6 to 5-8.

TRANSFER CYCLE TYPE	# (5)	MAX.(NSEC) ISA(1)	E-ISA(2)	EISA(3)
COMPATIBLE	1 2	558 308	533 293	560 (4) 320 (4)
TYPE "A"	1 2	NA NA	293 173	320 (4) 200 (4)
TYPE "B"	1 2	NA NA	173 53	190 (4) 70 (4)

TABLE 5-2: TC (INPUT TO PLATFORM) SIGNAL LINE DELAY FROM COMMAND SIGNAL LINE

NOTES:

(1) BCLK = 8.00 MHZ
(2) BCLK = 8.33 MHZ
(3) EISA bus specification BCLK = 8.33 MHZ
(4) See additional parameters in the EISA bus specification.
(5) Refer to numbers in Figures 5-3 through 5-8.

When programmed as an output, the TC signal line is valid relative to the end of the active period of the COMMAND signal line. The numbers are summarized in Table 5-3.

TRANSFER CYCLE TYPE	# (6)	ISA(1)	MIN.(NSEC) E-ISA(2)	EISA(3)
COMPATIBLE	6 7 8	60 525	60 500 700	60 500 700
TYPE "A"	6 7 7 8	NA NA NA NA	60 300 (4) 240 (5) 480	60 300 (4) 240 (5) 480
TYPE "B"	6 7 7 8	NA NA NA NA	-30 200 (4) 180 (5) 240	-30 200 (4) 180 (5) 240

TABLE 5-3: TC (OUTPUT FROM PLATFORM) SIGNAL LINE DELAY FROM COMMAND SIGNAL LINE

NOTES:

(1) BCLK = 8.00 MHZ
(2) BCLK = 8.33 MHZ
(3) EISA bus specification BCLK = 8.33 MHZ
(4) Read of I/O resource
(5) Write of I/O resource
(6) Refer to numbers in Figures 5-3 through 5-8.

> The IEEE P996 Specification specifies that Parameters 6 and 7 are 60 nanoseconds minimum.

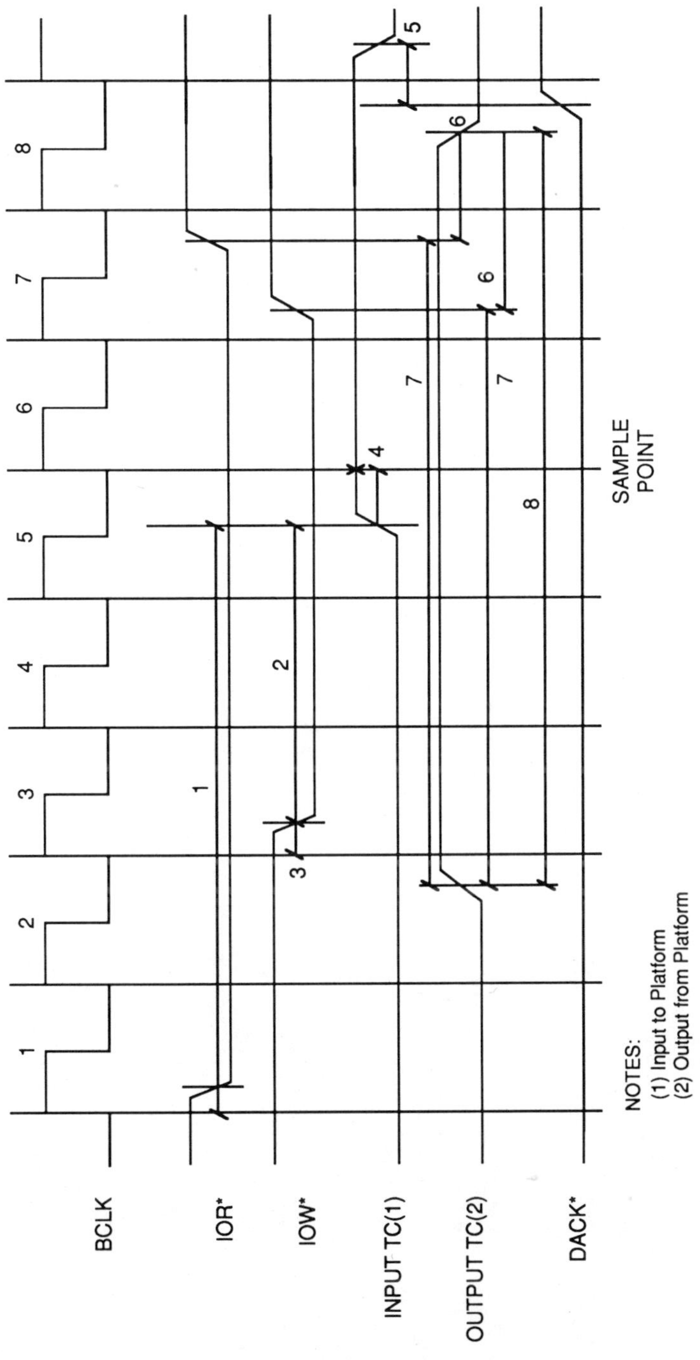

FIGURE 5-6: TERMINAL COUNT TIMING
COMPATIBLE TRANSFER CYCLE (STANDARD CYCLE)

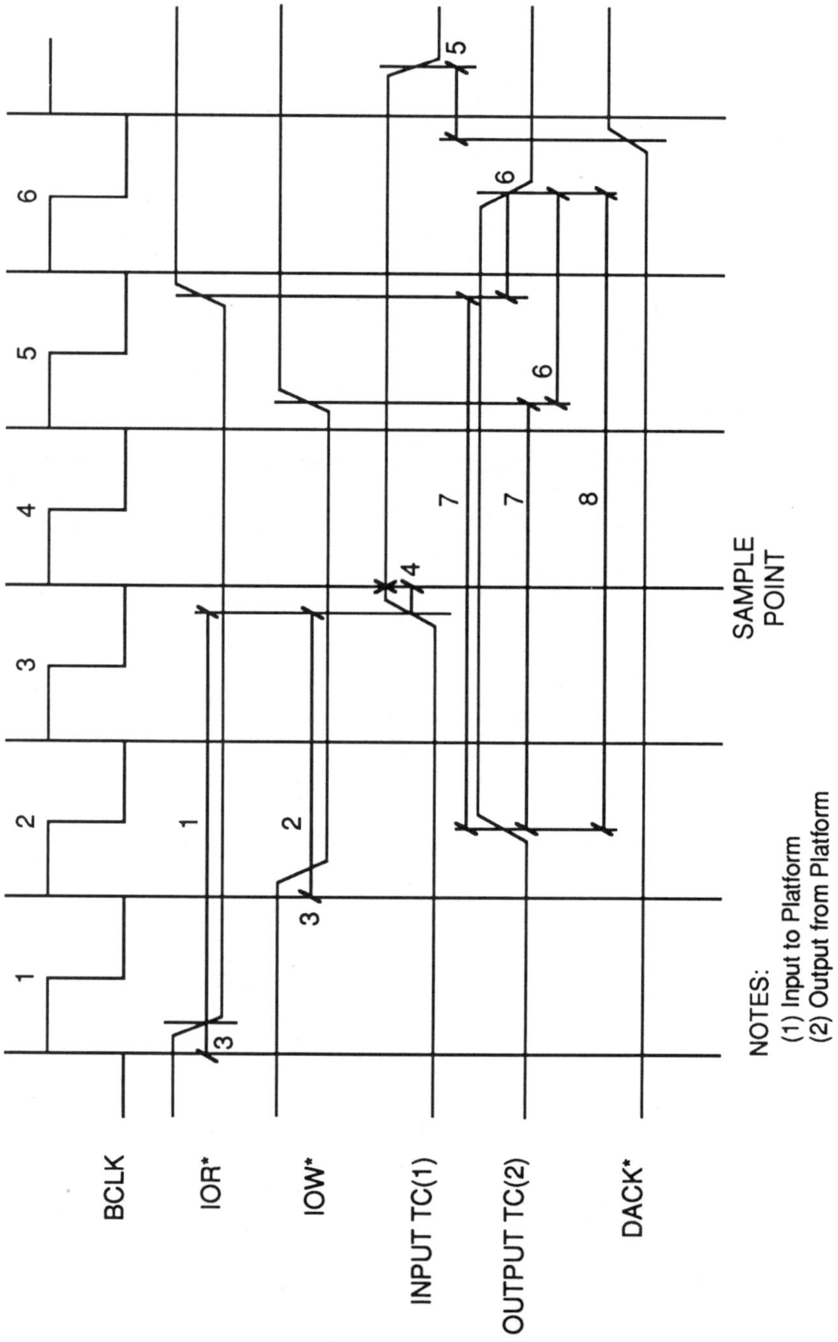

FIGURE 5-7: TERMINAL COUNT TIMING
TYPE "A" TRANSFER CYCLE (STANDARD CYCLE)

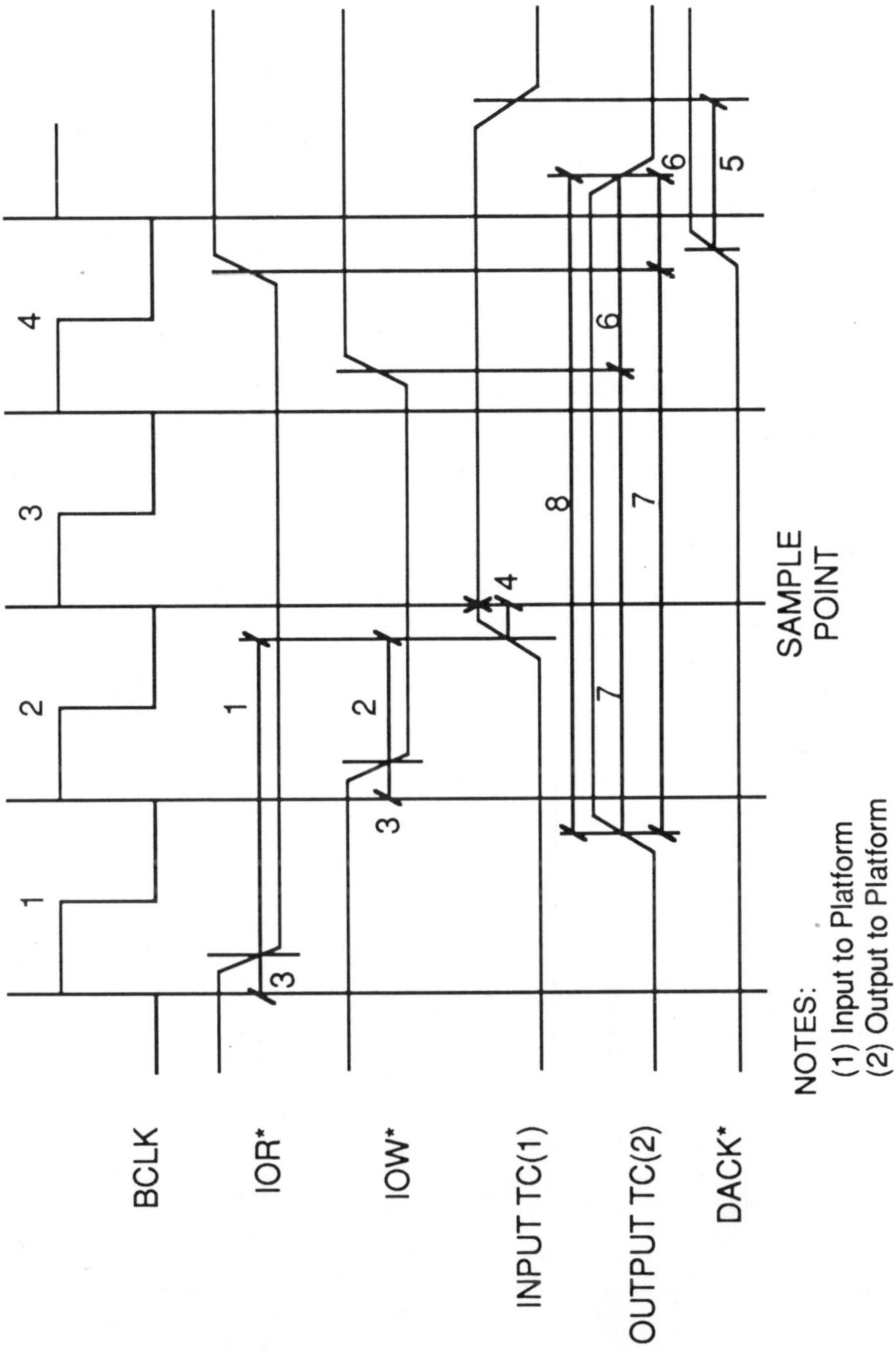

FIGURE 5-8: TERMINAL COUNT TIMING
TYPE "B" TRANSFER CYCLE (STANDARD CYCLE)

POWER SIGNALS

The power signal lines provide +5, -5, +12, -12 , and GND to the add-on cards. All of the power values are available on 8 bit slots. The 8/16 bit slots provide one additional +5 and one additional GND connection.

The current available for each voltage is outlined in Table 5-1. Please refer to Chapter 13 for further information concerning tolerance, noise, and other issues.

VOLTAGE	8 BIT	8/16 BIT
+5	3.0 AMPS	4.5 AMPS
+12	1.5 AMPS	1.5 AMPS
-5	1.5 AMPS	1.5 AMPS
-12	1.5 AMPS	1.5 AMPS

TABLE 5-4: CONNECTOR CURRENT

Not all platforms have sufficiently large power supplies to provide the maximum current to each slot.

CHAPTER SIX

BUS ACCESS CYCLES

INTRODUCTION

The access cycles support reading from and writing to memory and I/O resources by the bus owner. The two bus owners that can execute access cycles are the platform CPU and add-on bus owner cards. Table 3-1 in Chapter 3 outlines the support level of access cycles by the different bus owners and platforms.

As outlined in Chapter 3, there are three types of access cycles. The type is determined by the exact interpretation of signal lines and the length of the cycle. The length of the cycle is dependent on the cycle's data width and the logic level of the SRDY* and IOCHRDY signal lines. The shortest cycle for a given data width is the no wait state cycle. The longest cycle for a given data width is the ready cycle. When both the SRDY* and the IOCHRDY signal lines are active, a no wait state cycle is executed. When the SRDY* signal line is inactive and the IOCHRDY signal line is active, a standard cycle is executed. The length of the standard cycle varies with data width. Finally, when the IOCHRDY* signal line is inactive, a ready cycle is executed.

Most of the signal lines become active and inactive on the bus independent of the BCLK signal line. The only exception is the SRDY* signal line. From this viewpoint, the ISA bus is defined as an asynchronous bus. On an ISA platform, the CPU executes access cycles in an asynchronous fashion with a synchronized monitoring of the SRDY* signal line. The DMA, refresh, and add-on bus owner cards execute their respective cycles in an asynchronous fashion without supporting the SRDY* signal line.

The E-ISA platform executes ISA bus compatible cycles in the same fashion as the ISA platform. However, the E-ISA platform CPU uses an expanded interpretation of the no wait state cycle, and samples some bus signal lines relative to the BCLK signal line. The unique EISA(32)-only cycles require additional control lines not provided on the ISA connector; consequently, they will not be discussed in this book.

Figure 6-1 details the access cycle timing relationships for platform CPUs or add-on bus owners on ISA or E-ISA platforms. Table 6-1 lists the values of these timing relationships.

Table 6-1 has several key pieces of information. First, the "#/DET" (parameter number / detail) column relates the parameter number of the table to Figure 6-1. Second, it provides detailed information on how the number was derived. The timing numbers are affected by the settling time on the bus. The table is based on the default values for 8 slots as outlined in Chapter 10. The interpretation of the #/DET entries are shown on the following page.

The A.C. timing tables and signal waveform drawings in this section represent a composite of information from several sources in addition to the information that is original to this book. The following sources have provided portions of the A.C. timing and signal waveforms:

"ISA BUS SPECIFICATIONS and APPLICATION NOTES, January 30, 1990", Copyright 1989, Intel Corporation

"TECHNICAL REFERENCE GUIDE, EXTENDED INDUSTRY STANDARD ARCHITECTURE EXPANSION BUS", Copyright COMPAQ COMPUTER Corp., 1989

"PERSONAL COMPUTER BUS STANDARD P996" DRAFT D2.00, January 18, 1990, Copyright IEEE Inc., 1990.

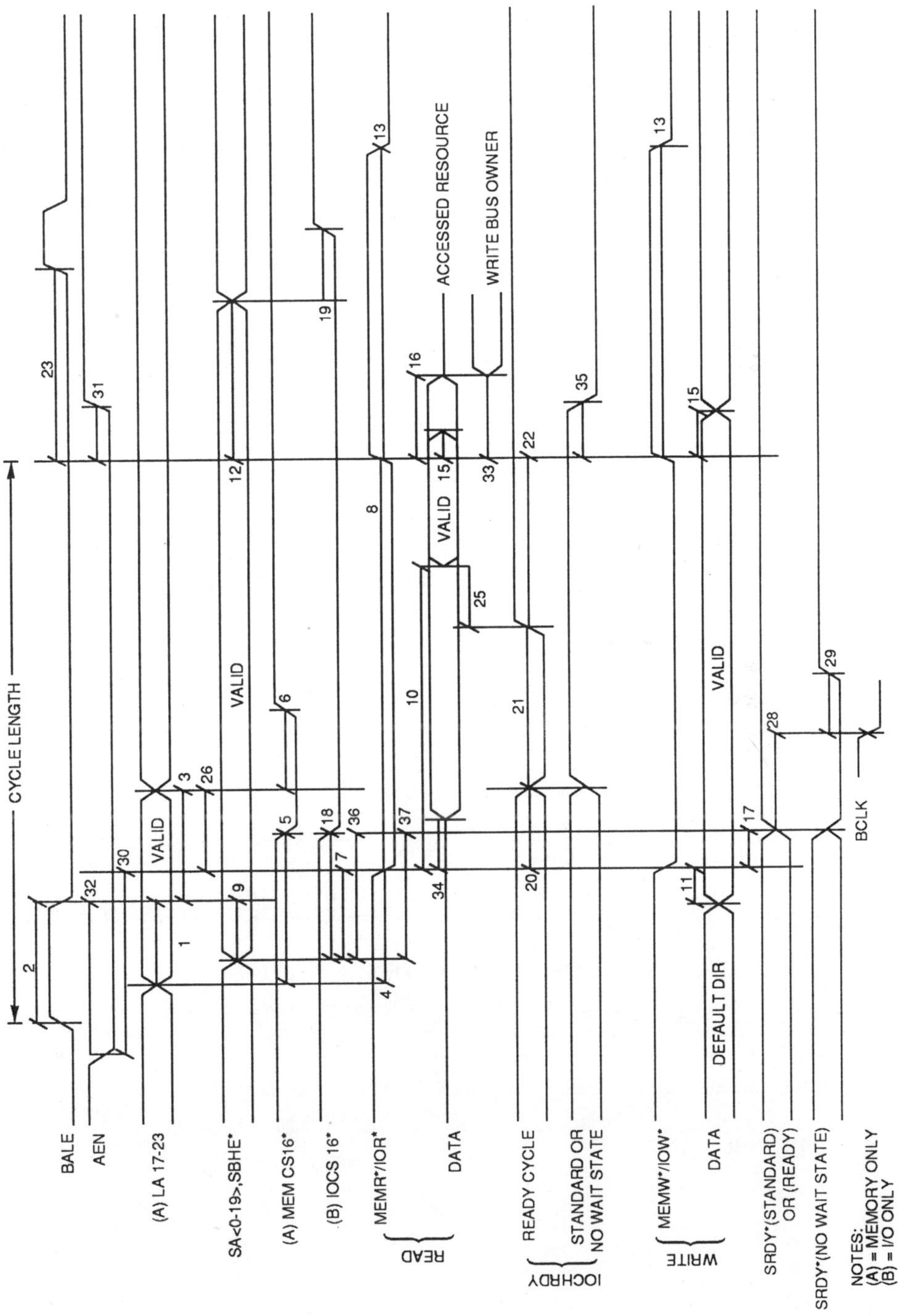

FIGURE 6-1: MEMORY AND I/O ACCESS CYCLES

DETAIL	INTERPRETATION (SEE CHAPTER 10 FOR MORE DETAILED INFORMATION)
#	NO SETTLING TIME INCLUDED.
#/A	SKEW OF TWO SIGNALS BETWEEN RESOURCES. ONE HAS A DELAY OF 0 NANOSECONDS, AND THE OTHER HAS A DELAY OF 11 NANOSECONDS.
#/B	SIGNAL IS DRIVEN BY ONE RESOURCE TO ANOTHER. THE SECOND RESOURCE MUST RESPOND BY DRIVING A SIGNAL BACK TO THE FIRST RESOURCE. # INCLUDES THE "ROUND TRIP" TIME OF 11 + 11 = 22 NANOSECONDS.
#/C	SAME AS "B", EXCEPT ON OF THE RESOURCES RELIES ON A 1K OHM PULL-UP RESISTOR. # INCLUDES THE "ROUND TRIP" TIME OF 11 + 78 = 89 NANOSECONDS.
#/D	SAME AS "B", EXCEPT ONE OF THE RESOURCES RELIES ON A 300 OHM PULL UP RESISTOR. # INCLUDES THE "ROUND TRIP" TIME OF 11 + 22 = 33 NANOSECONDS.
#/E	SAME AS "A", EXCEPT THE WORST CASE INTERPRETATION IS THAT BOTH SIGNALS ARE EITHER 0 NANOSECONDS OR 11 NANOSECONDS. THE NET RESULT IS A DIFFERENCE OF 0 NANOSECONDS.

"DRIVEN BY" AND "MEASURED BY" INTERPRETATION

ADD-ON	ADD-ON ACCESSED RESOURCE
ISA-B	ISA OR E-ISA COMPATIBLE ADD-ON CARD BUS OWNER
CPU	PLATFORM CPU
PLAT	PLATFORM CIRCUITRY

MIN. AND MAX. INTERPRETATION

ALL TIMES ARE IN NANOSECONDS WITH AN 8.00 MHZ BCLK.
THE NUMBER IN "[ ]" REFLECTS THE CALCULATED VALUE FOR AN 8.33 MHZ BCLK.

EISA CPU /MASTER INTERPRETATION

THE "EISA" ENTRIES ARE ACCORDING TO THE EISA BUS SPECIFICATION AND ARE MEASURED AT THE PLATFORM CPU, NOT AT THE ADD-ON CARD BEING ACCESSED. IF THE BUS OWNER IS THE ADD-ON CARD AND IS SYNCHRONIZED WITH BCLK, THE "EISA" ENTRIES ARE USED. THE NUMBERS FOLLOWING THE "/" ARE THE EISA BUS SPECIFICATIONS WHEN THE BUS OWNER IS AN ADD-ON CARD NOT SYNCHRONIZED WITH BCLK.

ALL NUMBERS ARE ROUNDED DOWNWARD FOR .5 OR BELOW.

SMEMR* AND SMEMW* HAVE A 10 NANOSECOND DELAY FROM MEMR* AND MEMW* WHEN A PLATFORM CPU, DMA CONTROLLER, OR REFRESH CONTROLLER IS BUS OWNER.

WHEN AN ADD-ON CARD IS BUS OWNER, THE DELAY IS 10 + 11 = 21 NANOSECONDS, BECAUSE THE PLATFORM CIRCUITRY DRIVES THE SIGNAL FROM MEMR* AND MEMW* (10 NANOSECONDS) AFTER RECEIVING IT FROM THE ADD-ON CARD (11 NANOSECONDS).

REF# /DET	DESCRIPTION OF EVENT	DRIVEN BY	MEASURED AT	MIN [8.33]	MAX	EISA CPU /MASTER
1/A	LA<17-23> VALID SETUP TO FALLING EDGE OF BALE	CPU	ADD-ON	100 [93]		116
2/A	BALE ACTIVE TO INACTIVE	PLAT	ADD-ON	50 [48]		30
3/A	LA<17-23> VALID HOLD FROM FALLING EDGE OF BALE	CPU	ADD-ON	15		22
4/A	LA<17-23> VALID SETUP TO MEMORY COMMAND ACTIVE (NOTE: FOR ISA-B USE #7) 8 BITS	CPU	ADD-ON	172 [162]		176
	16 BITS			109 [102]		112
5	MEMCS16* ACTIVE FROM VALID LA<17-23>	ADD-ON	ADD-ON		80 [72]	96/120
5/B		ADD-ON	CPU/ISA-B		102 [94]	
5	MEMCS16* INACTIVE FROM VALID LA<17-23>	ADD-ON	ADD-ON		66 [58]	96/120
5/D		ADD-ON	CPU/ISA-B		99 [91]	
6/E	MEMCS16* VALID HOLD FROM INVALID LA<17-23>	ADD-ON	ADD-ON	0		0/0
		ADD-ON	CPU/ISA-B	0		
7/A	SA<0-19> & SBHE* VALID TO MEMORY COMMAND ACTIVE (NOTE: FOR ISA-B USE THIS FOR LA<17-23>) 8 BITS	CPU /ISA-B	ADD-ON	91 [89]		88/120
	16 BITS			28 [23]		28/120
8/A	MEMORY COMMAND ACTIVE TO INACTIVE STANDARD 8 BITS	CPU /ISA-B	ADD-ON	530 [509]		530/530
	16 BITS			219 [209]		230/230
	NO WAIT STATE 8 BITS			154 [147]		166
	16 BITS			92 [87]		104
	I/O COMMAND ACTIVE TO INACTIVE STANDARD 8 BITS			530 [509]		530/530
	16 BITS			154 [147]		166/166

TABLE 6-1: ACCESS CYCLE TIMINGS

REF# /DET	DESCRIPTION OF EVENT	DRIVEN BY	MEASURED AT	MIN	MAX	EISA CPU /MASTER
				[8.33]		
	NO WAIT STATE 8 BIT			154 [147]		166
9/A	SA<0-19> & SBHE* VALID TO FALLING EDGE OF BALE	CPU	ADD-ON	29 [26]		28
10	VALID READ DATA FROM MEMORY COMMAND ACTIVE					
	STANDARD (SEE NOTE 1) 8 BITS	ADD-ON	ADD-ON		482 [460]	490/505
	16 BITS				173 [161]	194/194
	NO WAIT STATE 8 BITS				110 [103]	122
	16 BITS				48 [43]	70
10/B	VALID READ DATA FROM MEMORY COMMAND ACTIVE					
	(SEE NOTE 2) STANDARD 8 BITS	ADD-ON	CPU/ISA-B		504 [482]	
	16 BITS				195 [173]	
	NO WAIT STATE 8 BITS				132 [125]	
	16 BITS				70 [65]	
10	VALID READ DATA FROM I/O COMMAND ACTIVE					
	STANDARD 8 BITS	ADD-ON	ADD-ON		482 [462]	490/505
	16 BITS				110 [103]	130/130
	NO WAIT STATE 8 BITS				110 [103]	122
10/B	VALID READ DATA FROM I/O COMMAND ACTIVE					
	(SEE NOTE 1) STANDARD 8 BITS	ADD-ON	CPU/ISA-B		504 [484]	
	16 BITS				132 [125]	
	NO WAIT STATE 8 BITS				132 [125]	

TABLE 6-1: ACCESS CYCLE TIMINGS (CONTINUED)

REF# /DET	DESCRIPTION OF EVENT	DRIVEN BY	MEASURED AT	MIN [8.33]	MAX	EISA CPU /MASTER
11/A	VALID WRITE DATA SETUP TO MEMORY COMMAND ACTIVE NO WAIT STATE & STANDARD	CPU /ISA-B	ADD-ON			
	8 BITS EVEN			-4 [-9]		22/37
	ODD			-56 [-61]		22/37
	16 BITS			-45 [-47]		-40/-40
	VALID WRITE DATA SETUP TO I/O COMMAND ACTIVE STANDARD					
	8 BITS EVEN			-4 [-9]		22/37
	ODD			-56 [-61]		22/37
	16 BITS			22 [15]		22/37
	NO WAIT STATE 8 BITS EVEN			-4 [-9]		22
	ODD			-56 [-61]		22
12/A	SA<0-19> & SBHE* VALID HOLD FROM COMMAND INACTVE (<LA 17-23 FOR ISA-B)	CPU /ISA-B	ADD-ON	42 [40]		32/32
13/A	MEMORY COMMAND INACTIVE TO MEMORY COMMAND ACTIVE	CPU /ISA-B	ADD-ON			
	8 BITS			159 [152]		/170
	16 BITS			97 [92]		/170
	I/O COMMAND INACTIVE TO I/O COMMAND ACTIVE	CPU /ISA-B	ADD-ON	159 [152]		/170
15	READ DATA VALID HOLD FROM INACTIVE MEMORY OR I/O COMMAND	ADD-ON	ADD-ON	0 [0]		0/0
15/E	WRITE DATA VALID HOLD FROM INACTIVE MEMORY COMMAND	CPU /ISA-B	ADD-ON			
	(SEE NOTE 3) 8 BITS			25 [25]		9/25
	16 BITS			25 [25]		25/25
15/E	WRITE DATA VALID HOLD FROM INACTIVE I/O COMMAND	CPU	ADD-ON	25 [25]		25/25

TABLE 6-1: ACCESS CYCLE TIMINGS (CONTINUED)

REF# /DET	DESCRIPTION OF EVENT	DRIVEN BY	MEASURED AT	MIN [8.33]	MAX [8.33]	EISA CPU /MASTER
16	READ COMMAND INACTIVE TO SD<0-15> TRISTATE	ADD-ON	ADD-ON		30 [30]	30/30
16/E	WRITE COMMAND INACTIVE TO SD<0-15> TRISTATE	CPU /ISA-B	ADD-ON		30 [30]	30/30
17	SRDY* VALID FROM COMMAND ACTIVE	ADD-ON	ADD-ON			
	8 BITS				72 [67]	80
	16 BITS				10 [5]	15
	8 BITS	ADD-ON	CPU		94 [89]	
	16 BITS				32 [27]	
18	IOCS16* ACTIVE FROM VALID SA<0-15>	ADD-ON	ADD-ON		100 [95]	160/160
18/B		ADD-ON	CPU/ISA-B		122 [117]	
18	IOCS16* INACTIVE FROM VALID SA<0-15>	ADD-ON	ADD-ON		74 [69]	160/160
18/D		ADD-ON	CPU/ISA-B		96 [91]	
19/A	IOCS16* VALID HOLD FROM VALID SA<0-15>	ADD-ON	ADD-ON	0 [0]		0/0
20[¥]	IOCHRDY INACTIVE FROM ACTIVE COMMAND	ADD-ON	ADD-ON			
	8 BITS				362 [344]	398/398
	16 BITS				59 [54]	86/86
20/B[£]	8 BITS	ADD-ON	CPU/ISA-B		384 [366]	I/O= 80/80
	16 BITS				81 [76]	
20[β]	IOCHRDY ACTIVE FROM ACTIVE COMMAND	ADD-ON	ADD-ON			
	8 BITS				362 [344]	398/398
	16 BITS				59 [54]	86/86
20/C[√]	8 BITS	ADD-ON	CPU/ISA-B		451 [433]	I/O= 80/80
	16 BITS				148 [143]	

TABLE 6-1: ACCESS CYCLE TIMINGS (CONTINUED)

Special notes below apply to early IBM PC/AT 16 Bit I/O:

¥ - May not be possible because of Bus settling time.
£ - 16 bit I/O may be as low as 15 ns.
β - May not be possible because of Bus R/C rise time.
√ - 16 bit I/O may be as low as 0 ns.

REF# /DET	DESCRIPTION OF EVENT	DRIVEN BY	MEASURED AT	MIN [8.33]	MAX [8.33]	EISA CPU /MASTER
21	IOCHRDY INACTIVE PULSE (SEE NOTE 4)	ADD-ON	ADD-ON	125 [120]	15.6 MICROSECONDS	
22/E	COMMAND ACTIVE HOLD FROM IOCHRDY ACTIVE	CPU /ISA-B	ADD-ON	125 [120]		116/120
23	BALE ACTIVE FROM COMMAND INACTIVE	CPU /ISA-B	ADD-ON	35 [33]		
24	BCLK PERIOD	PLAT.	ADD-ON	125 [120]		120
25	VALID READ DATA FROM IOCHRDY ACTIVE	ADD-ON	ADD-ON			
	8 BITS EVEN				74 [69]	70
	ODD				64 [59]	70
	16 BITS				74 [69]	80
25/A	8 BITS EVEN	ADD-ON	CPU/ISA-B		85 [80]	/70
	ODD				75 [70]	/70
	16 BITS				85 [80]	/80
26/A	LA<17-23> INVALID FROM COMMAND ACTIVE	CPU /ISA-B	ADD-ON			
	8 BITS			-32 [-32]		-30
	16 BITS			30 [28]		25
28	SRDY* VALID SETUP TO THE FALLING EDGE OF BCLK	ADD-ON	ADD-ON	22 [22]		10
28/B		ADD-ON	CPU	0 [0]		
29/A	SRDY* VALID HOLD FROM THE FALLING EDGE OF BCLK	ADD-ON	ADD-ON	22 [22]		20
30/A	AEN VALID TO I/O COMMAND ACTIVE	PLAT.	ADD-ON	100 [100]		100
31/A	AEN VALID FROM I/O COMMAND INACTIVE	PLAT.	ADD-ON	30 [30]		30
32/A	AEN VALID TO FALLING EDGE OF BALE	PLAT.	ADD-ON	100 [100]		100
33	DATA DRIVEN IN WRITE DIRECTION FROM INACTIVE READ COMMAND	CPU /ISA-B	CPU/ISA-B	30 [30]		

TABLE 6-1: ACCESS CYCLE TIMINGS (CONTINUED)

REF# /DET	DESCRIPTION OF EVENT	DRIVEN BY	MEASURED AT	MIN [8.33]	MAX	EISA CPU /MASTER
34	DATA DRIVEN FROM READ COMMAND ACTIVE	ADD-ON	ADD-ON	0 [0]		0/0
35	IOCHRDY ACTIVE HOLD FROM COMMAND INACTIVE	ADD-ON	ADD-ON	0 [0]		
36	LA<17-23> VALID TO SRDY* ACTIVE	CPU	ADD-ON			
	8 BITS			283 [268]		280
	16 BITS			158 [148]		156
36/B	8 BITS	CPU	CPU	305 [290]		
	16 BITS			180 [170]		
37	SA<0-19> & SBHE* VALID TO SRDY* ACTIVE	CPU	ADD-ON			
	8 BITS			188 [178]		192
	16 BITS			61 [56]		68
37/B	8 BITS	CPU	CPU	210 [200]		
	16 BITS			83 [78]		

TABLE 6-1: ACCESS CYCLE TIMINGS (CONTINUED)

NOTES:

(1) Parameter 10 in the above table represents the numbers computed for an actual IBM platform. However, to meet the same timings when the same add-on memory card is part of an 8 bit transfer cycle, the actual time must be less. According to Parameter 24 of Table 7-1, the numbers 482 [460] should be replaced by 325 [310]. Indeed, this appears to be what was done by the EISA bus specification.

(2) Parameter 10/B has the same situation as explained for Parameter 10 above. According to Parameter 24/B of Table 7-1 the numbers 504 [482] should be replaced by 347 [332].

(3) The EISA bus specification requires only a 9 nanosecond minimum data hold time for 8 bits. This probably is measured relative to the SMEMW signal line. The entry in the table is measured relative to the MEMW* signal line. See Chapters 5 and 10 for more information concerning the relationship of the SMEMW*, SMEMR*, MEMW*, and MEMR* signal lines.*

(4) The IBM Technical Reference Manual specifies 2.5 microseconds.

The following sections will first discuss the details for each ISA bus cycle executed in an ISA platform, and will then cover the enhancements for the E-ISA platform.

STANDARD ACCESS MEMORY CYCLE (ISA PLATFORM)

The standard access memory cycle can be executed by the platform CPU or by an add-on bus owner card. (See Figures 6-2 and 6-3.) When the platform CPU is the bus owner, the functional portion of the cycle begins by pulsing the BALE signal line to indicate that the LA, SA, AEN, and SBHE* signal lines are valid. For an add-on bus owner card, the functional portion of the cycle begins when the COMMAND signal line is driven active. When the platform CPU is bus owner, the overall cycle is a standard cycle if the IOCHRDY signal line is active within a specific time relative to the COMMAND signal line becoming active, and if the SRDY* signal line is inactive at the center falling edge of the second clock period of the BCLK cycle. When the add-on bus owner card owns the bus, a standard cycle is dependent only upon the IOCHRDY signal line being active within a specific time relative to the COMMAND signal line becoming active.

For a 16 bit access cycle, the bus owner must request a 16 bit cycle and the accessed resource must support a 16 bit cycle. For an 8 bit access cycle, either the bus owner must request an 8 bit cycle or the accessed resource must be an 8 bit resource. This is summarized in Table 6-2.

BUS OWNER SBHE*	BUS OWNER SA0	BUS OWNER DATA SIZE	ACCESSED RESOURCE MEMCS16*	DATA SIZE OF CYCLE
INACT.	0	8	X	8
ACT.	1	8	X	8
ACT.	0	16	INACT.	8
ACT.	0	16	ACT.	16

TABLE 6-2: ISA & E-ISA PLATFORM MEMORY CYCLE DATA SIZE

Once the data size is established, the bus owner and the accessed resource know the minimum length of time that the COMMAND signal line is driven active. For a read cycle, the data from the accessed resource must be valid within a maximum amount of time from the MEMR* (SMEMR*) signal line becoming active. For a write cycle, the data must be valid prior to the MEMW* (SMEMW*) signal line becoming active. The data must remain valid a minimum amount of time after the appropriate command line is driven inactive. The COMMAND signal line becoming inactive indicates the completion of the cycle.

> *When the bus owner executes an access cycle, the SMEMR* and SMEMW* signal lines must be monitored by certain memory resources. If either line is active, then only the first 1 megabyte of the memory address space is being accessed. For a memory resource in an 8/16 bit slot, the LA<20-23> signal lines are available to indicate the first 1 megabyte. Also, the MEMR* and MEMW* signal lines are available. For a memory resource in an 8 bit slot, the LA<20-23>, MEMR*, and MEMW* signal lines are not available. Since the SMEMR* and SMEMW* signal lines are the only memory command signal lines available in an 8 bit slot, 8 bit memory resources can only reside in the first 1 megabyte of the address space.*

Of particular note is the requirement for back to back bus cycles. There is a minimum time from a COMMAND signal line becoming inactive to the next active COMMAND or BALE signal line. If the present cycle is 16 bits and the next is 8 bits, the 16 bit minimum time between active COMMAND and BALE signal lines applies. If the present cycle is 8 bits and the next is 16 bits, the 8 bit minimum time between active COMMAND and BALE signal lines applies.

The length of the standard access memory cycle will vary depending on the data width of the cycle. The reference point for the beginning of the cycle will vary for different bus owners. Table 6-3 summarizes the cycle length in terms of TCLK (period of BCLK). Due to bus settling times and device delays,

TCLK should be considered as approximate for the "CYCLE BEGINS" column.

BUS OWNER	CYCLE BEGINS	CYCLE ENDS	CYCLE SIZE	CYCLE LENGTH
PLAT FORM CPU	.5 TCLK TO ACT. BALE	END COMMAND ACT. PERIOD	8 BITS 16 BITS	6 TCLK 3 TCLK
ADD- ON	1.5 TCLK TO ACT. COMMAND	END COMMAND ACT. PERIOD	8 BITS	6 TCLK
ADD- ON	1 TCLK TO ACT. COMMAND	END COMMAND ACT. PERIOD	16 BITS	3 TCLK

TABLE 6-3: ISA & E-ISA PLATFORM STANDARD ACCESS MEMORY CYCLE LENGTH

STANDARD ACCESS MEMORY CYCLE (E-ISA PLATFORM)

When an add-on bus owner card is executing the cycle, the operation of a standard access memory cycle is defined for the E-ISA platform in the same way as for an ISA platform. (See Figures 6-2 and 6-3.)

The standard access memory cycle is also defined by an E-ISA platform in the same way as an ISA platform when the platform CPU is executing the cycle, except that the sampling of the IOCHRDY signal line has been enhanced. (See Figures 6-2 and 6-3.) For a 16 bit cycle, the INITIAL SAMPLING WINDOW for the IOCHRDY signal line is the third BCLK cycle. For an 8 bit cycle, the INITIAL SAMPLING WINDOW for the IOCHRDY signal line is the sixth BCLK cycle. For the cycle to remain a standard cycle, the IOCHRDY signal line must be sampled active during the sampling window.

See Tables 6-2 and 6-3 in the STANDARD ACCESS MEMORY CYCLE (ISA PLATFORM) section for cycle data size and cycle length information, respectively. Also see this section for back to back bus cycle information.

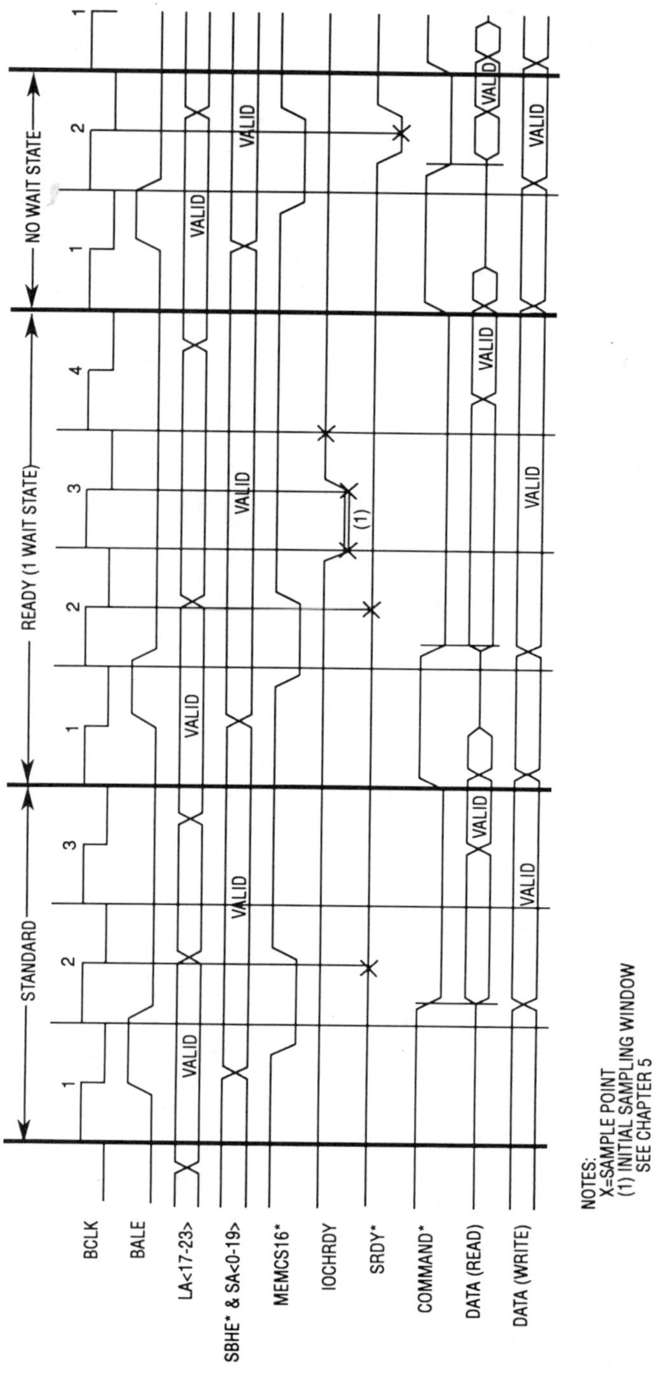

FIGURE 6-2-A: 16 BIT MEMORY ACCESS CYCLES PLATFORM CPU

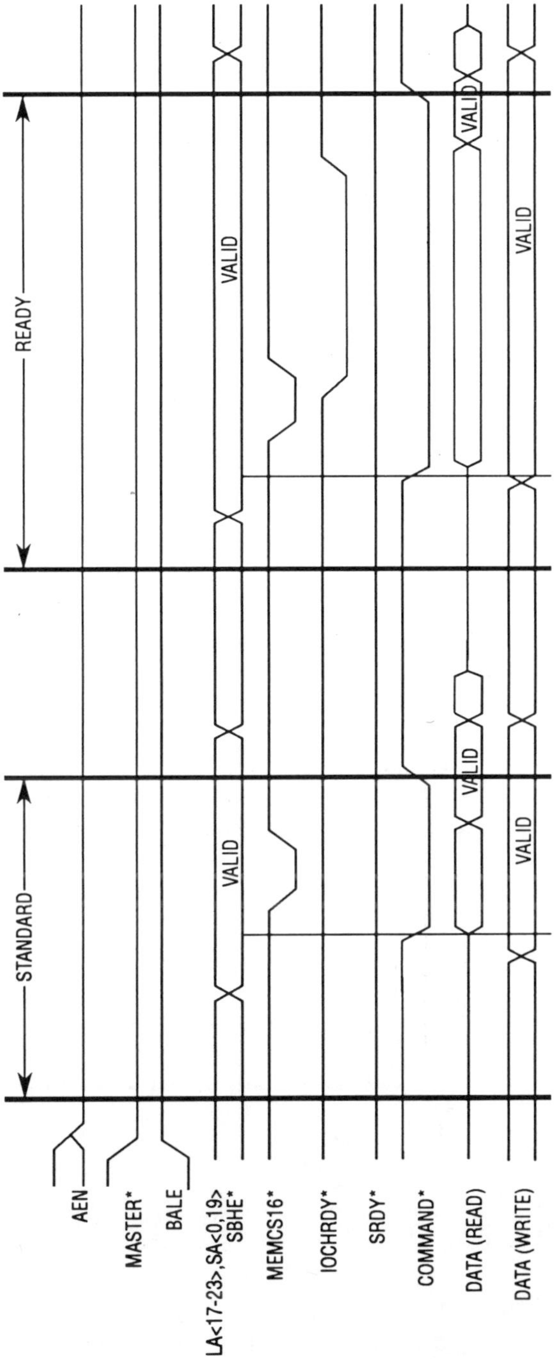

FIGURE 6-2-B: 16 BIT MEMORY ACCESS CYCLES
ADD-ON BUS OWNER CARD

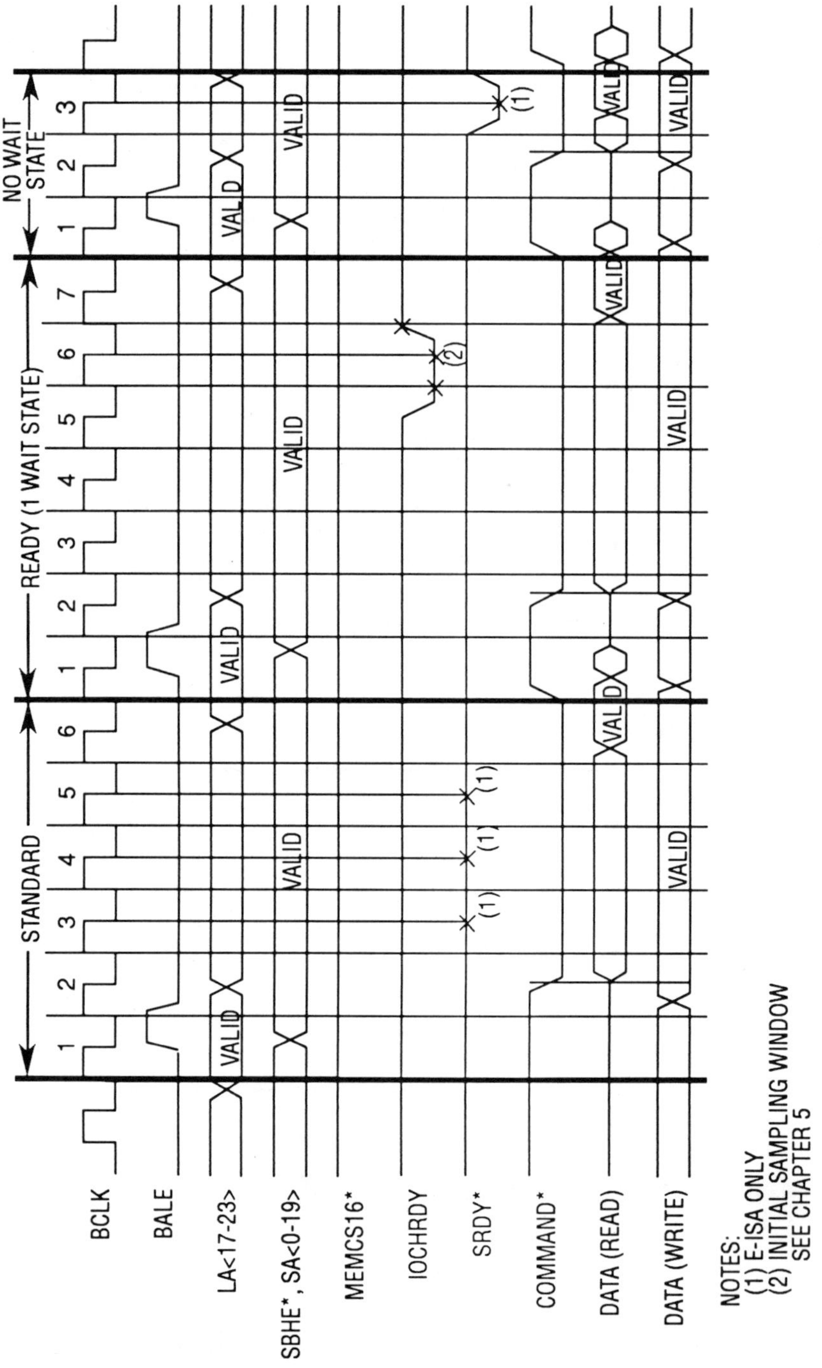

FIGURE 6-3-A: 8 BIT MEMORY ACCESS CYCLES PLATFORM CPU

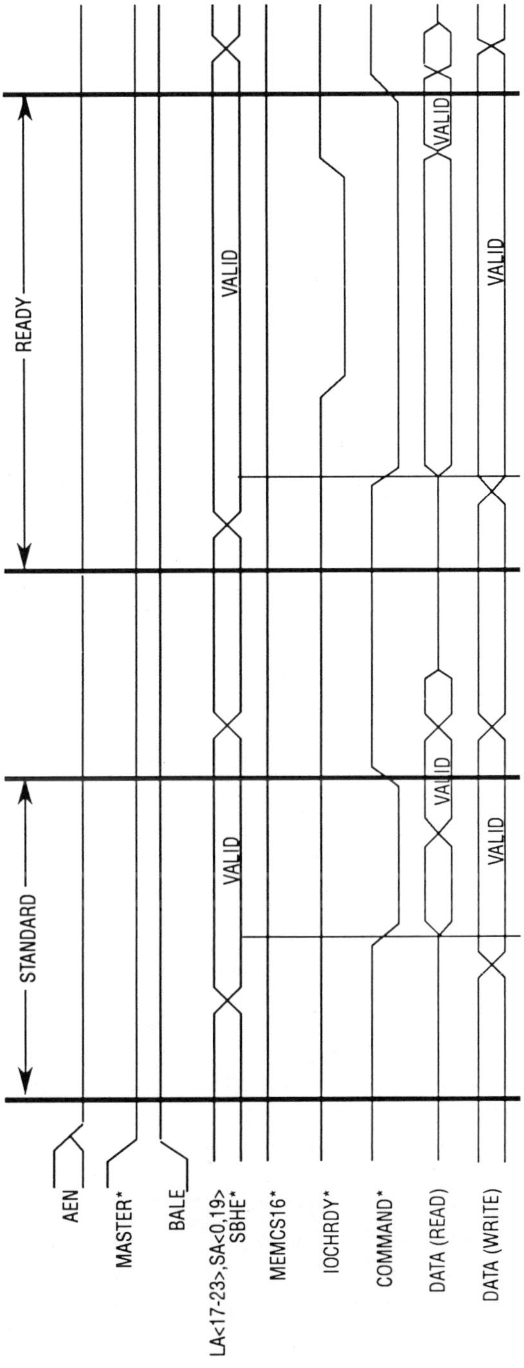

FIGURE 6-3-B: 8 BIT MEMORY ACCESS CYCLES
ADD-ON BUS OWNER CARD

STANDARD ACCESS I/O CYCLE (ISA PLATFORM)

The standard access I/O cycle can be executed by either a platform CPU or an add-on bus owner card. (See Figures 6-4 and 6-5.) The standard access I/O cycle operates in the same way as a standard memory cycle, except that the LA<17-23> and the SA<16-19> signal lines are not used. When the platform CPU is the bus owner, the functional portion of the cycle begins by pulsing the BALE signal line to indicate that SA, AEN, and SBHE* signal lines are valid. For an add-on bus owner card, the functional portion of the cycle begins when a COMMAND signal line is driven active. When the platform CPU is bus owner, the overall cycle timings are those of a standard cycle if the IOCHRDY signal line is active and the SRDY* signal line is inactive at the center falling edge of the second clock period of the BCLK cycle. When the add-on bus owner card owns the bus, the cycle is a standard cycle dependent only on the IOCHRDY signal line being active within a specific time of the COMMAND signal line becoming active. For a 16 bit access cycle, the bus owner must request a 16 bit cycle and the accessed resource must support a 16 bit cycle. For an 8 bit access cycle, either the bus owner must request an 8 bit cycle or the accessed resource must be only an 8 bit resource. This is summarized in Table 6-4.

BUS OWNER			ACCESSED RESOURCE	DATA SIZE
SBHE*	SA0	DATA SIZE	IOCS16*	OF CYCLE
INACT.	0	8	X	8
ACT.	1	8	X	8
ACT.	0	16	INACT.	8
ACT.	0	16	ACT.	16

TABLE 6-4: ISA & E-ISA PLATFORM ACCESS I/O CYCLE DATA SIZE

Once the data size is established, the bus owner and the accessed resource know the minimum length of time that the COMMAND signal line is driven active. For a read cycle, the data from the accessed resource must be valid within a maximum amount of time from IOR* signal line becoming active. For a write cycle, the data must be valid prior to the IOW* signal line becoming active. The data must remain valid for a minimum amount of time after the appropriate

COMMAND line is driven inactive. The COMMAND signal line becoming inactive indicates the completion of the cycle.

Of particular note is the requirement for back to back bus cycles. There is a requirement for a minimum time from the COMMAND signal line becoming inactive to the next COMMAND or BALE signal line becoming active. If the present cycle is 16 bits and next is 8 bits, then the 16 bit minimum time applies. If the present cycle is 8 bits and the next is 16 bits, then the 8 bit minimum time applies.

The length of the standard I/O memory cycle will vary depending on the data width for the cycle. The reference point for the beginning of the cycle varies for different bus owners. Table 6-5 summarizes the cycle length in terms of TCLK (period of BCLK). Due to bus settling times and device delays, TCLK should be considered as approximate for the "CYCLE BEGINS" column.

BUS OWNER	CYCLE BEGINS	CYCLE ENDS	CYCLE SIZE	CYCLE LENGTH
PLATFORM CPU	.5 TCLK TO ACT. BALE	END COMMAND ACT. PERIOD	8 BITS 16 BITS	6 TCLK 3 TCLK
ADD-ON	1.5 TCLK TO ACT. COMMAND	END COMMAND ACT. PERIOD	8 BITS 16 BITS	6 TCLK 3 TCLK

TABLE 6-5: ISA & E-ISA PLATFORM STANDARD ACCESS I/O CYCLE LENGTH

On an ISA platform, only the SA<0-9> signal lines are driven valid by the bus owner during an I/O access cycle.

STANDARD ACCESS I/O CYCLE (E-ISA PLATFORM)

When an add-on bus owner card is executing the cycle, the operation of a standard access I/O cycle is defined for an E-ISA platform in the same way as for an ISA platform. (See Figures 6-4 and 6-5.)

When the platform CPU is executing the cycle, the operation of a standard access I/O cycle is defined by an E-ISA

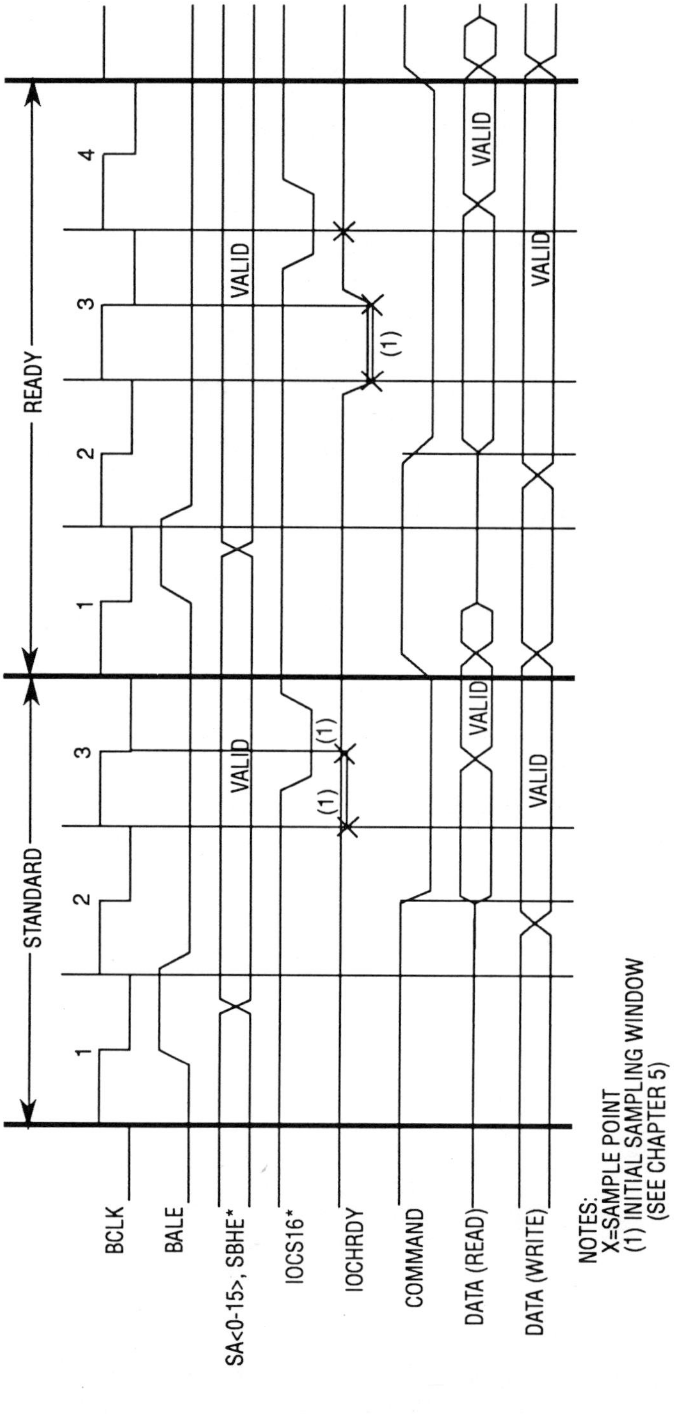

FIGURE 6-4-A: 16 BIT I/O ACCESS CYCLES
PLATFORM CPU

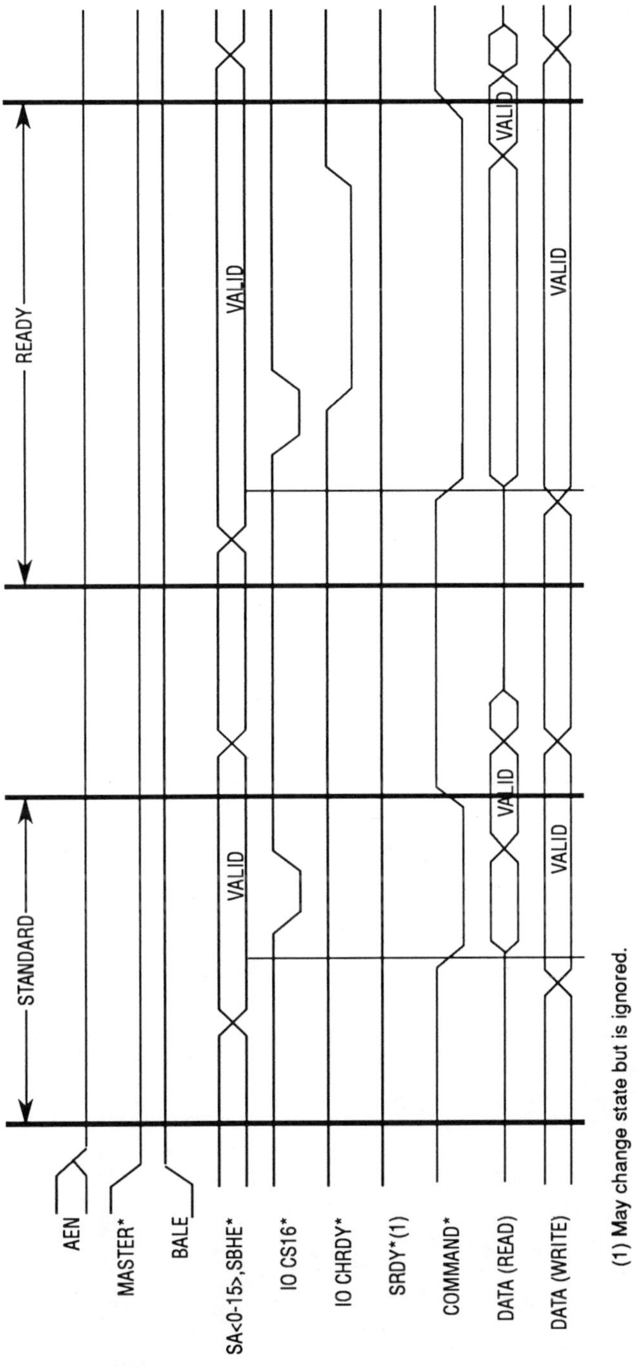

FIGURE 6-4-B: 16 BIT I/O ACCESS CYCLES
ADD-ON BUS OWNER CARD

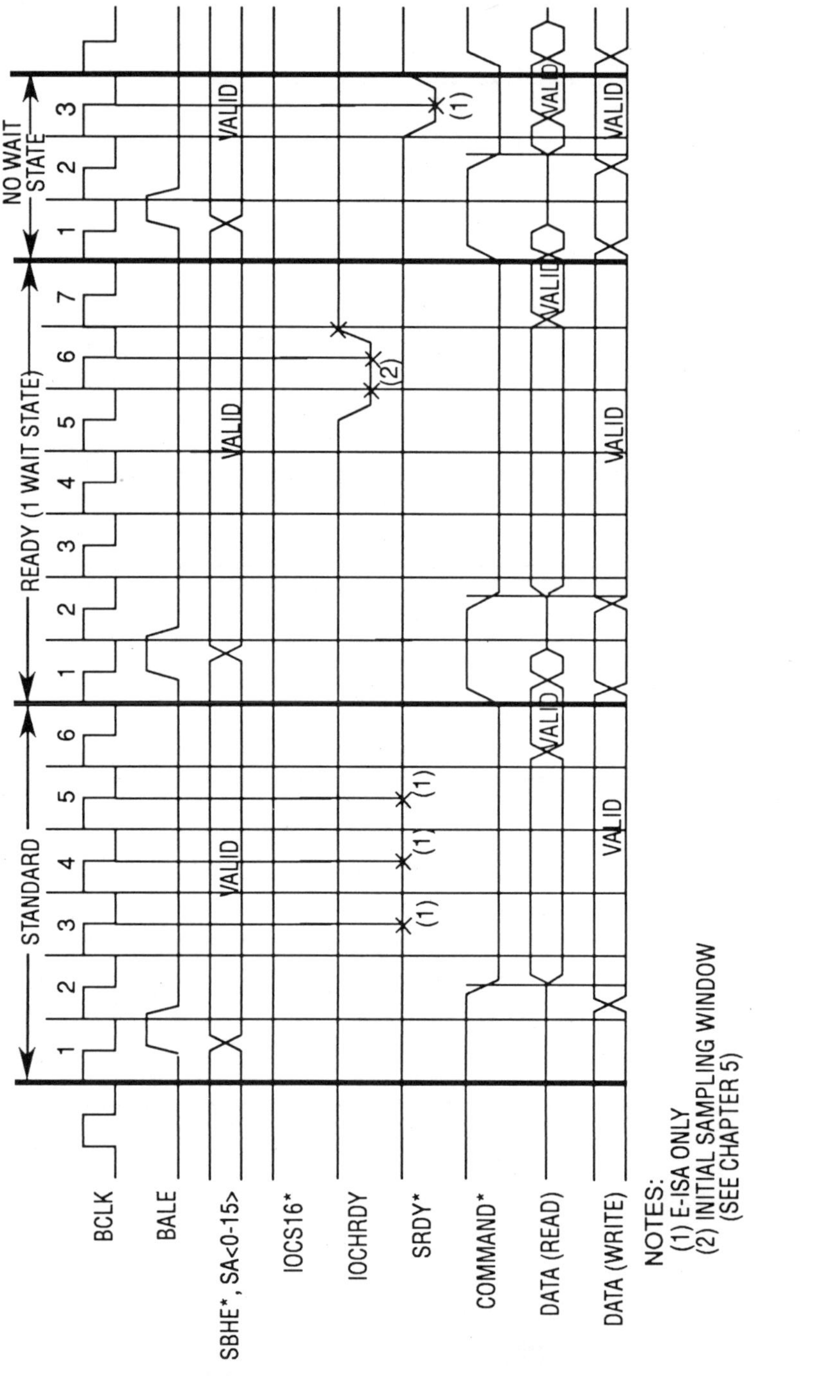

FIGURE 6-5-A: 8 BIT I/O ACCESS CYCLES
PLATFORM CPU

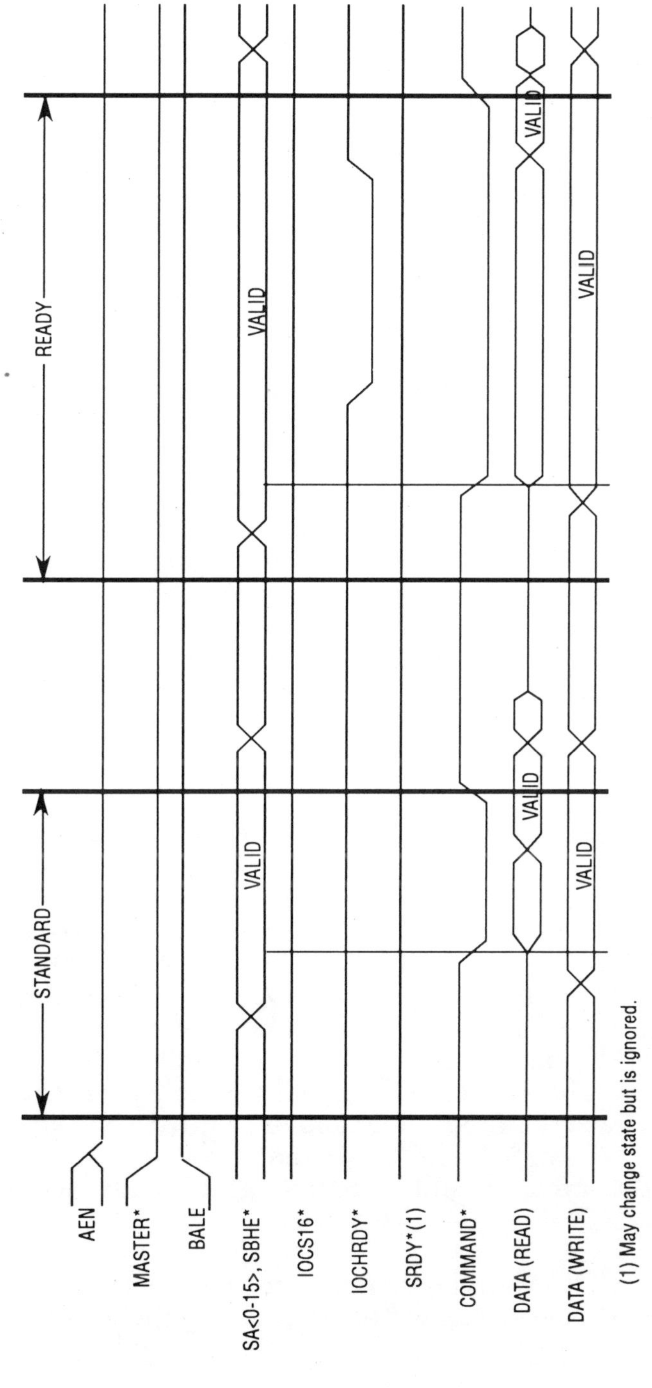

FIGURE 6-5-B: 8 BIT I/O ACCESS CYCLES
ADD-ON BUS OWNER CARD

platform in the same way as an ISA platform, except that the sampling of the IOCHRDY signal line has been enhanced. (See Figures 6-4 and 6-5.) For a 16 bit cycle, the INITIAL SAMPLING WINDOW for the IOCHRDY signal line is the third BCLK cycle. For an 8 bit cycle, the INITIAL SAMPLING WINDOW for the IOCHRDY signal line is the sixth BCLK cycle. For the cycle to remain a standard cycle, the IOCHRDY signal must be sampled active during the sampling window.

See Tables 6-4 and 6-5 in the STANDARD ACCESS I/O CYCLE (ISA PLATFORM) section for cycle data size and cycle length information, respectively. Also see this section for back to back bus cycle information.

> *On an E-ISA platform, during an I/O access cycle, the platform CPU is only required to drive the SA<0-11> signal lines valid. The value of the SA<12-15> signal lines do not have to be valid, in that the AENx signal lines "encode" the correct value. An I/O access cycle executed by an add-on bus owner card must drive the SA<0-15> signal lines valid. The SA<12-15> signal lines must be driven valid so the platform circuitry can "encode" their value into the AENx signal lines.*

READY ACCESS MEMORY CYCLE (ISA PLATFORM)

On an ISA platform, the ready access memory cycle can be executed by the platform CPU or add-on bus owner card. (See Figures 6-2 and 6-3.) The cycle begins in the same fashion as the standard access memory cycle. The standard access memory cycle becomes a ready access memory cycle if the IOCHRDY signal line is driven inactive by the accessed resource. The IOCHRDY signal line must be driven inactive within a "specific time" of an active COMMAND signal line. (If the SRDY* signal line is driven active during this "specific time", the cycle is still executed as a ready access memory cycle.)

The COMMAND signal line remains active until the IOCHRDY signal line is driven active by the accessed resource.

During a memory read cycle, the IOCHRDY signal line becoming active indicates to the bus owner that valid data is available. During a write cycle, the IOCHRDY signal line becoming active indicates to the bus owner that the accessed resource has accepted the data. For either read or write cycles, the activation of the IOCHRDY signal line is the indication to the bus owner to complete the cycle. The cycle is completed when the COMMAND signal line becomes inactive.

See Table 6-2 in the STANDARD ACCESS MEMORY CYCLE (ISA PLATFORM) section for cycle data size information. Also see this section for back to back bus cycle, SMEMR*, and SMEMW* signal line information.

The length of the ready access memory cycle depends on when the IOCHRDY signal line is reactivated. The ready cycle length is extended in increments of the BCLK period. (In order to insure refresh of memory, the longest inactive period allowed for the IOCHRDY signal line is 15.6 microseconds.) The minimum time for a ready access memory cycle is controlled by several factors: First, once the IOCHRDY signal line is reactivated, the COMMAND signal line will not be deactivated until after a minimum of one BCLK period has elapsed. Second, the bus owner must have sufficient sampling time for the IOCHRDY signal line; consequently, the minimum inactive period of the signal is one BCLK period. Finally, sufficient decode time from a COMMAND signal line going active must be allocated. These requirements collectively establish the minimum length of a ready access memory cycle. The minimum cycle length, in terms of TCLK (period of BCLK), is summarized in Table 6-6. Due to bus settling times and device delays, TCLK should be considered as approximate for the "CYCLE BEGINS" column.

BUS OWNER	CYCLE BEGINS	CYCLE ENDS	CYCLE SIZE	CYCLE LENGTH
PLATFORM CPU	.5 TCLK TO ACT. BALE	END COMMAND ACT. PERIOD	8 BITS 16 BITS	7 TCLK 4 TCLK
ADD-ON	1.5 TCLK TO ACT. COMMAND	END COMMAND ACT. PERIOD	8 BITS	7 TCLK
ADD-ON	1 TCLK TO ACT. COMMAND	END COMMAND ACT. PERIOD	16 BITS	4 TCLK

TABLE 6-6: ISA & E-ISA PLATFORM MINIMUM READY ACCESS MEMORY CYCLE LENGTH

READY ACCESS MEMORY CYCLE (E-ISA PLATFORM)

When an add-on bus owner card executes the cycle, the operation of a ready access memory cycle is defined by an E-ISA platform in the same way as for an ISA platform. (See Figures 6-2 and 6-3.) Refer to the READY ACCESS MEMORY CYCLE (ISA PLATFORM) section for further information. If the add-on bus owner card wants to reference the IOCHRDY signal line to the BCLK signal line, the bus cycle must resemble a platform CPU's cycle. See the SPECIAL E-ISA NOTE section in Chapter 1 for more information.

When the platform CPU is executing the cycle, the operation of a ready access memory cycle is also defined by an E-ISA platform in the same way as an ISA platform, except the sampling of the IOCHRDY signal line has been enhanced. (See Figures 6-2 and 6-3.) For a 16 bit cycle, the INITIAL SAMPLING WINDOW for the IOCHRDY signal line is the third BCLK cycle. For an 8 bit cycle, the INITIAL SAMPLING WINDOW for the IOCHRDY signal line is the sixth BCLK cycle. For the cycle to become a ready cycle, the IOCHRDY signal line must be sampled inactive during the sampling window. The IOCHRDY signal line is then sampled at the rising edge of each subsequent BCLK cycle. The cycle is completed at the end of the BCLK cycle in which the IOCHRDY signal line was sampled active. See the IOCHRDY signal line section of Chapter 5 for more information. The purpose of this enhanced sampling scheme is to give E-ISA platforms a performance edge over the ISA platforms.

See Table 6-2 in the STANDARD ACCESS MEMORY CYCLE (ISA PLATFORM) section for cycle data size information. Also see this section for back to back bus cycle, SMEMR*, and SMEMW* signal line information. See Table 6-6 in the READY ACCESS MEMORY CYCLE (ISA PLATFORM) section for cycle length information.

READY ACCESS I/O CYCLE (ISA PLATFORM)

On an ISA platform, the ready access I/O cycle can be executed by either the platform CPU or an add-on bus owner card. (See Figures 6-4 and 6-5.) The cycle begins in the same fashion as the standard access I/O cycle. The standard access I/O cycle becomes a ready access I/O cycle if the IOCHRDY signal line is driven inactive by the accessed resource. The IOCHRDY signal line must be driven inactive within a "specific time" of an active COMMAND signal line. (If the SRDY* signal line is driven active during this "specific time", then the cycle is still executed as a ready access I/O cycle.)

The COMMAND signal line remains active until the IOCHRDY signal line is driven active by the accessed resource. During an I/O read cycle, the IOCHRDY signal line becoming active indicates to the bus owner that valid data is available. During a write cycle, the IOCHRDY signal line becoming active indicates to the bus owner that the accessed resource has accepted the data. For either read or write cycles, the activation of the IOCHRDY signal line is the indication to the bus owner to complete the cycle. The cycle is completed when the COMMAND signal line becomes inactive.

See Table 6-4 in the STANDARD ACCESS I/O CYCLE (ISA PLATFORM) section for cycle data width information. Also see this section for back to back bus cycle information.

The length of the ready access I/O cycle depends on when the IOCHRDY signal line is reactivated. The ready cycle length is extended in increments of the BCLK signal. (In order to insure memory refresh, the longest inactive period of the IOCHRDY signal line is 15.6 microseconds.) The minimum time for a ready access I/O cycle is controlled by several factors. First, once the IOCHRDY signal line is reactivated, the COMMAND signal line will not be deactivated until after a minimum of one BCLK period has elapsed. Second, the bus owner must have sufficient sampling time for the IOCHRDY signal line. Consequently, the minimum inactive period of the signal is one BCLK period. Finally, sufficient decode time from a COMMAND signal line going active must be allocated. These requirements collectively establish the minimum length of a ready access I/O cycle. The minimum cycle length, in terms of TCLK (period of BCLK), is summarized in Table 6-7. Due to bus settling times and device delays, TCLK should be considered as approximate for the "CYCLE BEGINS" column.

BUS OWNER	CYCLE BEGINS	CYCLE ENDS	CYCLE SIZE	CYCLE LENGTH
PLATFORM CPU	.5 TCLK TO ACT. BALE	END COMMAND ACT. PERIOD	8 BITS 16 BITS	7 TCLK 4 TCLK
ADD-ON	1.5 TCLK TO ACT. COMMAND	END COMMAND ACT. PERIOD	8 BITS	7 TCLK
ADD-ON	1.5 TCLK TO ACT. COMMAND	END COMMAND ACT. PERIOD	16 BITS	4 TCLK

TABLE 6-7: ISA & E-ISA PLATFORM MINIMUM READY ACCESS I/O CYCLE LENGTH

READY ACCESS I/O CYCLE (E-ISA PLATFORM)

When an add-on bus owner card executes the cycle, the operation of a ready access I/O cycle is defined for an E-ISA platform in the same way as for an ISA platform. (See Figures 6-4 and 6-5.) Please refer to the READY ACCESS I/O CYCLE (ISA PLATFORM) section for further information. If the add-on bus owner card wants to reference the IOCHRDY signal line to the BCLK signal line, the bus cycle must resemble a platform CPU's cycle. See the SPECIAL E-ISA NOTE section in Chapter 1 for more information.

When the platform CPU is executing the cycle, the operation of a ready access I/O cycle is also defined by an E-ISA platform in the same way as an ISA platform except that the sampling of the IOCHRDY signal line has been enhanced. (See Figures 6-4 and 6-5.) For a 16 bit cycle, the INITIAL

SAMPLING WINDOW for the IOCHRDY signal line is the third BCLK cycle. For an 8 bit cycle, the INITIAL SAMPLING WINDOW for the IOCHRDY signal line is the sixth BCLK cycle. For the cycle to become a ready cycle, the IOCHRDY signal line must be sampled inactive during the sampling window. The IOCHRDY signal line is then sampled at the rising edge of each subsequent BCLK cycle. The cycle is completed at the end of the cycle where the IOCHRDY signal line was sampled active. See the IOCHRDY signal line section of Chapter 5 for more information. The purpose of this enhanced sampling scheme is to give E-ISA platforms a performance edge over the ISA platforms.

See Table 6-4 in the STANDARD ACCESS I/O CYCLE (ISA PLATFORM) section for cycle data size. Also see this section for back to back bus cycle information. See Table 6-7 in the READY ACCESS I/O CYCLE (ISA PLATFORM) section for cycle length information.

NO WAIT STATE ACCESS MEMORY CYCLE (ISA PLATFORM)

On an ISA platform, the no wait state memory access cycle can only be executed when the platform CPU is accessing a 16 bit memory resource. (See Figure 6-2.) The cycle begins in the same fashion as the standard access memory cycle. The standard access memory cycle becomes a no wait state access memory cycle if the SRDY* signal line is driven active by the accessed resource. The SRDY* signal line must be driven active within a "specific time" of an active COMMAND signal line relative to the center falling edge of the second cycle of the BCLK cycle. The cycle will be completed at the end of the same cycle. The SRDY* signal line for a 16 bit memory access is sampled only at one point of the cycle. Sampling at other points is not needed because other cycle lengths are either a standard cycle or a ready cycle. (If the IOCHRDY signal line is driven inactive during this "specific time", then the cycle is executed as a ready access memory cycle.)

The completion of the no wait state access memory cycle is the same as the standard cycle, except that the active period of the COMMAND signal line is shorter.

See Table 6-2 in the STANDARD ACCESS MEMORY CYCLE (ISA PLATFORM) section for cycle data width information. Also see this section for back to back bus cycle information.

The length of the no wait state access memory cycle is defined for the platform CPU accessing a 16 bit memory resource. Table 6-8 summarizes the cycle length in terms of TCLK (period of BCLK). Due to bus settling times and device delays, TCLK should be considered as approximate for the "CYCLE BEGINS" column.

BUS OWNER /RESOURCE	CYCLE BEGINS	CYCLE ENDS	CYCLE SIZE	CYCLE LENGTH
PLATFORM CPU	.5 TCLK TO ACT. BALE	END COMMAND ACT. PERIOD	16 BITS	2 TCLK

TABLE 6-8: ISA PLATFORM NO WAIT STATE MEMORY ACCESS CYCLE LENGTH

> *The original IBM PC and XT platforms defined the SRDY* signal line connector pin as a card select response from slot 8. IBM AT platforms define this pin as the SRDY* signal line. For purposes of this document, it is assumed not to be supported on ISA platforms. The EISA bus specification CLEARLY DEFINES 8 BIT NO WAIT STATE CYCLES. Please see the NO WAIT STATE MEMORY ACCESS CYCLE (E-ISA PLATFORM) for more information. Please see the NOTE in the description of the SRDY* signal line section of Chapter 5 for more detailed information.*

NO WAIT STATE ACCESS MEMORY CYCLE (E-ISA PLATFORM)

The platform CPU executes the no wait state cycle on an E-ISA platform in the same fashion as an ISA platform CPU for

16 bit memory resources. (See Figure 6-2.) See the NO WAIT STATE ACCESS MEMORY CYCLE (ISA PLATFORM) section for more information. If an add-on bus owner card wants to execute a no wait state cycle, the bus cycle must resemble a platform CPU's cycle. See the SPECIAL E-ISA NOTE section in Chapter 1 for more information.

The EISA bus specification has expanded the definition of the no wait state memory access cycle. An E-ISA platform CPU not only supports accesses to 16 bit memory, but also supports accesses to 8 bit memory resources. (See Figure 6-3.) The cycle begins in the same fashion as the standard access memory cycle. The standard access memory cycle becomes a no wait state access cycle if the SRDY* signal line is driven active by the accessed resource. The SRDY* signal line must be driven active within a "specific time" of an active COMMAND signal line relative to the center falling edge of the BCLK signal line. The cycle will be completed at the end of the same cycle. The SRDY* signal line for an 8 bit memory access is sampled at three points of the cycle. The multiple sampling allows the cycle to complete earlier than a standard length cycle. Sampling at other points for a 16 bit cycle is not needed because other cycle lengths are either the standard cycle or a ready cycle.

> *The EISA bus specification requires the sampling of the SRDY* signal line at the falling edge of the center transition of the second cycle of the BCLK signal line for 16 bit cycles. The 8 bit COMMAND signal line is just becoming active at the center falling edge of the second cycle of the BCLK signal line. Consequently, for an 8 bit cycle, the first valid sampling is in the third cycle.*

The completion of the no wait state access memory cycle is the same as the standard memory cycle, except that the active period of the COMMAND signal line is shorter.

See Table 6-2 in the STANDARD ACCESS MEMORY CYCLE (ISA PLATFORM) section for data width information. Also see this section for back to back bus cycle, SMEMR*, and SMEMW* signal line information.

The length of the no wait state access memory cycle is defined for the platform CPU accessing an 8 or 16 bit memory resource. Table 6-9 summarizes the cycle length in terms of TCLK (period of BCLK). Due to bus settling times and device delays, the TCLK should be considered as approximate for the "CYCLE BEGINS" column.

BUS OWNER /RESOURCE	CYCLE BEGINS	CYCLE ENDS	CYCLE SIZE	CYCLE LENGTH
PLATFORM CPU	.5 TCLK TO ACTIVE BALE	END COMMAND	8 BITS	3 TCLK
	.5 TCLK TO ACTIVE BALE	END COMMAND ACT. PERIOD	16 BITS	2 TCLK

TABLE 6-9: E-ISA PLATFORM NO WAIT STATE MEMORY ACCESS CYCLE LENGTH

Due to bus settling issues, an add-on bus owner card cannot execute no wait state cycles. See Chapter 10 for more details.

Please see the NOTE in the description of the SRDY* signal line section of Chapter 5 for more detailed information.

NO WAIT STATE ACCESS I/O CYCLE (ISA PLATFORM)

The original IBM PC, XT, AT, and the ISA platforms do not support no wait state 16 bit I/O access cycles. The I/O command is active later in the cycle than a memory command. (See Figures 6-4 and 6-5.) The sampling of the SRDY* signal line would have to occur during the third BCLK signal line cycle instead of the second. However, the sampling in the third cycle is useless because the standard cycle length is three cycles. The standard cycle length of an 8 bit I/O access cycle is 6 cycles; consequently, defining no wait state cycles is possible. However, historically these platforms do not support no wait state 8 bit I/O access cycles.

NO WAIT STATE ACCESS I/O CYCLE (E-ISA PLATFORM)

On an E-ISA platform, the no wait state access I/O cycle can only be executed when the platform CPU is accessing an 8 bit resource. (See Figure 6-5.) The cycle begins in the same fashion as the standard access I/O cycle. The standard access I/O cycle becomes a no wait state access I/O cycle if the SRDY* signal line is driven active by the accessed resource. The SRDY* signal line must be driven active within a "specific time" of an active COMMAND signal line relative to the center falling edge of the third cycle of the BCLK cycle. The cycle will be completed at the end of the same cycle. The SRDY* signal line for an 8 bit access is sampled at three points of the cycle. The multiple sampling allows the cycle to complete earlier than a standard length cycle. (If the IOCHRDY signal line is driven inactive during this "specific time", then the cycle is executed as ready access memory cycle.) If an add-on bus owner card wants to execute a no wait state cycle, the bus cycle must resemble a platform CPU's cycle. See the SPECIAL E-ISA NOTE section in Chapter 1 for more information.

> *The EISA bus specification allows the first sampling of the SRDY* signal line at the center falling edge of the third clock cycle. The sampling cannot occur sooner because the 8 bit COMMAND signal line is becoming active only at the beginning of the second cycle of the BCLK signal line.*

The completion of the no wait state access I/O cycle is the same as the standard cycle, except that the active period of the COMMAND line is shorter.

See Table 6-4 in the STANDARD ACCESS I/O CYCLE (ISA PLATFORM) section for cycle data size. Also see this section for back to back bus cycle information.

The length of the no wait state access I/O cycle is defined for the platform CPU accessing an 8 bit I/O resource. Table 6-10 summarizes the cycle length in terms of TCLK (period of BCLK). Due to bus settling times and device delays TCLK should be considered as approximate for the "CYCLE BEGINS" column.

BUS OWNER /RESOURCE	CYCLE BEGINS	CYCLE ENDS	CYCLE SIZE	CYCLE LENGTH
PLATFORM CPU	.5 TCLK TO ACTIVE BALE	END COMMAND	8 BITS	3 TCLK

TABLE 6-10: E-ISA PLATFORM NO WAIT STATE I/O ACCESS CYCLE LENGTH

D… CLES

INTRODUCT…

The DM… …uted by the platform DMA contr… …n memory and I/O resources.

On an IS… …l lines during transfer cycles beco… …endent of the BCLK signal. O… …e, COMPATIBLE, is supported …

On an … …he signal lines (except for the IO… …es) become active and inactive i… …gnal line. The E-ISA platform e… …en ISA compatible add-on cards i… platform. However, the E-ISA pl… …tional types of transfer cycles bet… I/O cards and platform memory … …tional types of transfer cycles are…

Each … …er platform comes in two versions: … …fference between the two versions … …hich is controlled through the use …

The… transfer cycles: fly-by and non fly-… …a is read from the source at the sam… …stination. The data in a fly-

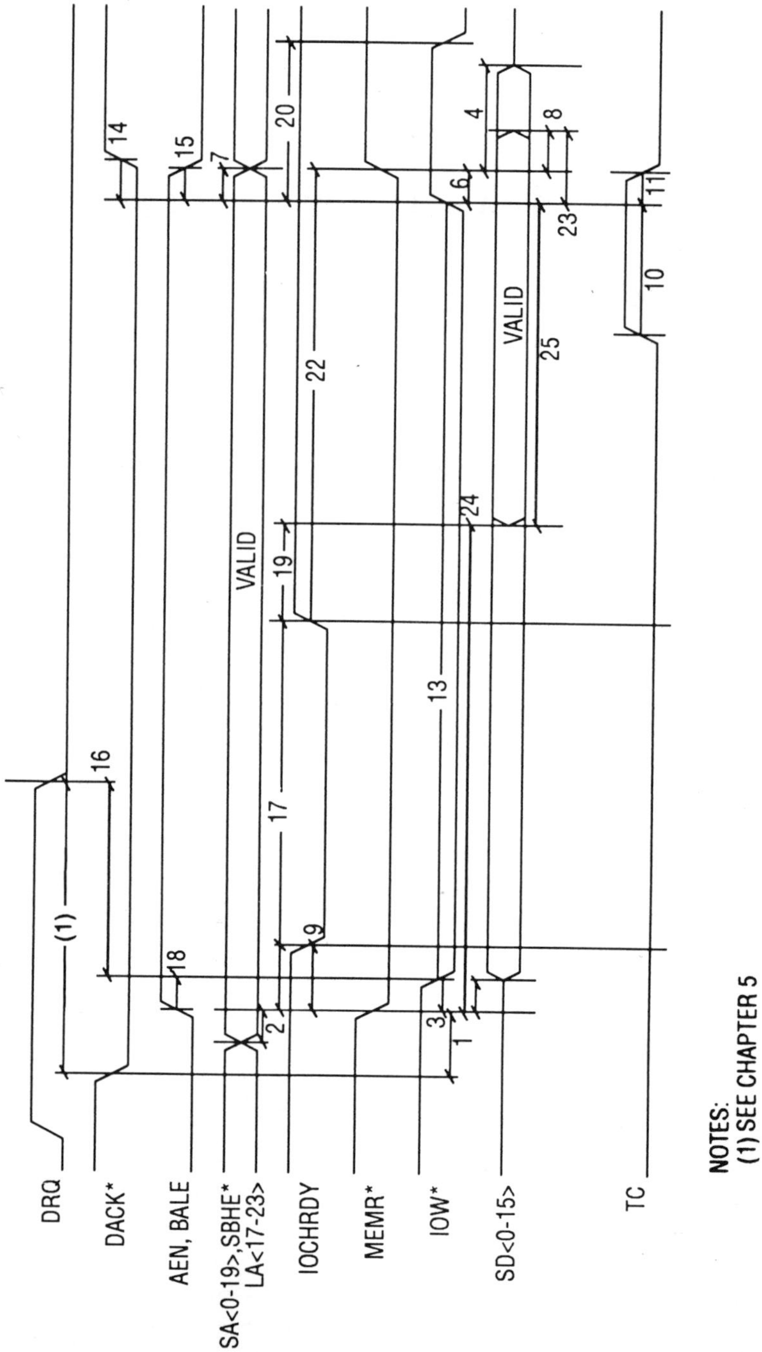

FIGURE 7-1: MEMORY READ - I/O WRITE TRANSFER CYCLE

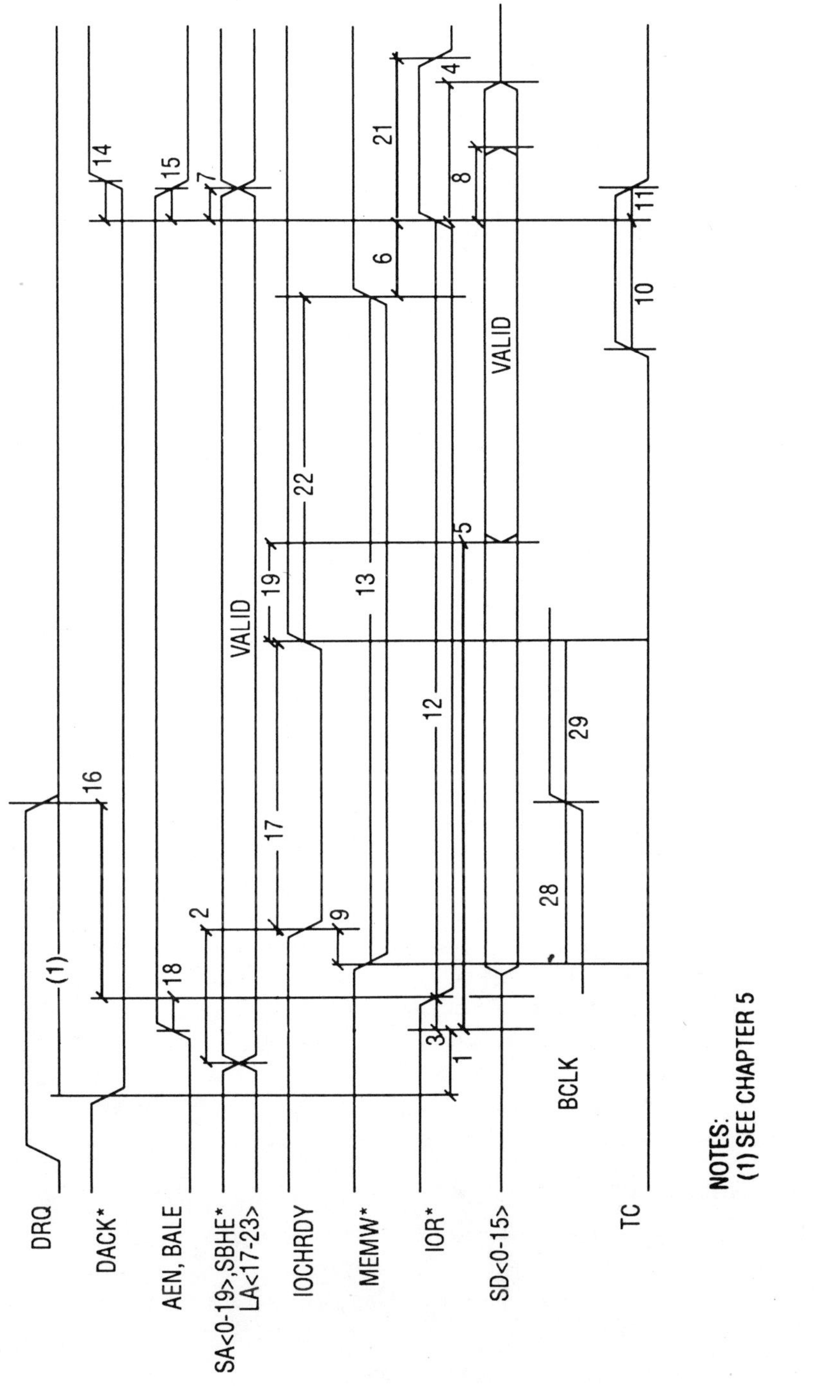

FIGURE 7-2: I/O READ - MEMORY WRITE TRANSFER CYCLE

by operation is never held or modified by the DMA controller. The fly-by transfer cycle only supplies a single value on the ADDRESS signal lines during each transfer cycle. This address is only valid for the memory resource of the cycle. The I/O resources receive only an active DACK* signal line to indicate which I/O resource participates in the cycle. This is in contrast to a non fly-by transfer where data is read from the source by the DMA controller. Subsequent to the data read cycle, the DMA controller executes a separate cycle to write data to the destination. Between the two cycles, the DMA controller holds the data and can view or modify it. The two individual cycles allow separate valid values on the ADDRESS signal lines; consequently, both memory and I/O can be addressed. The ISA and E-ISA platforms only execute the fly-by transfer cycles.

On an ISA platform, as outlined in Chapter 4, specific DACK* signal lines are defined for 8 bit I/O resources, and others are specific for 16 bit resources. The memory resource for a transfer cycle with an 8 bit I/O resource can be either 8 or 16 bits in width. The platform byte swapper circuitry will swap the bytes between high and low DATA signal lines as needed. For an 8 bit transfer, a 16 bit memory resource is required to read and write individual bytes from either the high or low bytes of the DATA signal lines. The memory resource for a 16 bit I/O resource must always be a 16 bit resource. This requirement simplifies the I/O resource circuitry because the requirement to read and write individual bytes on the I/O resource is removed.

The E-ISA platform has expanded the transfer cycles so that each channel can be programmed as an 8 or 16 bit I/O resource. The transfer cycles between ISA compatible I/O and memory add-on cards are the same as for the ISA platform. For transfer cycles between ISA compatible I/O resources and E-ISA platform memory resources, any of the three possible transfer cycles can be executed.

This chapter will explain the different versions and types of transfer cycles for ISA and E-ISA platforms. The basic transfer cycle is focused on the I/O resource. It is assumed that the memory resource operates in the same fashion during a transfer cycle as it does during an access cycle. Thus, all of the memory parameters listed in Table 7-1 are from Table 6-1. They are reprinted here for clarity and convenience. Any ISA compatible add-on memory slave card on an ISA or E-ISA platform can only participate in COMPATIBLE transfer cycles. If an ISA compatible add-on memory slave card participates in the other types of transfer cycles on an E-ISA or EISA(32) platform, the transfer cycle will appear to the add-on slave card as an ISA access cycle executed by an ISA add-on bus owner card.

Table 3-2 summarizes the transfers between the different resources. Figures 7-1 and 7-2 detail the transfer cycle timing relationships. Table 7-1 lists the values of these timing relationships.

Table 7-1 has several key pieces of information. First, the "#/DET" (parameter number / detail) column relates the parameter number of the table to Figures 7-1 and 7-2. Second, it defines the detail information for how the number was derived. The timing numbers are affected by the settling time on the bus. The table is based on the default values for 8 slots as outlined in chapter 10. The interpretation of the #/DET entries are shown on the following page.

The A.C. timing tables and signal waveform drawings in this section represent a composite of information from several sources in addition to the information that is original to this book. The following sources have provided portions of the A.C. timing and signal waveforms:

"ISA BUS SPECIFICATIONS and APPLICATION NOTES, January 30, 1990", Copyright 1989, Intel Corporation

"TECHNICAL REFERENCE GUIDE, EXTENDED INDUSTRY STANDARD ARCHITECTURE EXPANSION BUS", Copyright COMPAQ COMPUTER Corp., 1989

"PERSONAL COMPUTER BUS STANDARD P996" DRAFT D2.00 January 18, 1990, Copyright IEEE Inc., 1990.

DETAIL	INTERPRETATION
#	NO SETTLING TIME INCLUDED.
#/A	SKEW OF TWO SIGNALS BETWEEN RESOURCES. ONE HAS A DELAY OF 0 NANOSECONDS, AND THE OTHER HAS A DELAY OF 11 NANOSECONDS.
#/B	SIGNAL IS DRIVEN BY ONE RESOURCE TO ANOTHER. THE SECOND RESOURCE MUST RESPOND BY DRIVING A SIGNAL BACK TO THE FIRST RESOURCE. # INCLUDES THE "ROUND TRIP" TIME OF 11 + 11 = 22 NANOSECONDS.
#/C	SAME AS "B", EXCEPT ONE OF THE RESOURCES RELIES ON A 1K OHM PULL-UP RESISTOR. # INCLUDES THE "ROUND TRIP" TIME OF 11 + 78 = 89 NANOSECONDS.
#/D	SAME AS "B", EXCEPT ONE OF THE RESOURCES RELIES ON A 300 OHM PULL-UP RESISTOR. # INCLUDES THE "ROUND TRIP" TIME OF 11 + 22 = 33 NANOSECONDS.
#/E	SAME AS "A", EXCEPT THE WORST CASE INTERPRETATION IS THAT BOTH SIGNALS ARE EITHER 0 NANOSECONDS OR 11 NANOSECONDS. THE NET RESULT IS A DIFFERENCE OF 0 NANOSECONDS.

"DRIVEN BY" AND "MEASURED BY" INTERPRETATION

SOURCE	MEASURED AT THE ADD-ON CARD THAT PROVIDES THE DATA.
DESTIN	MEASURED AT THE ADD-ON CARD THAT ACCEPTS THE DATA.
DMA	PLATFORM DMA CONTROLLER.

MIN. AND MAX. INTERPRETATION

ALL TIMES ARE IN NANOSECONDS WITH AN 8.00 MHZ BCLK.
THE NUMBER IN "[ ]" REFLECTS THE CALCULATED VALUE FOR AN 8.33 MHZ BCLK.

EISA

THE ENTRIES ARE THOSE IN THE EISA BUS SPECIFICATION FOR THE PLATFORM DMA CONTROLLER.

SMEMR* AND SMEMW* HAVE A 10 NANOSECOND DELAY FROM MEMR* AND MEMW*.

REF# /DET	DESCRIPTION OF EVENT	DRIVEN BY	MEASURED AT	ISA MIN [8.33]	ISA MAX [8.33]	COMP. (3) MIN	COMP. (3) MAX	TYPE "A"(3) MIN	TYPE "A"(3) MAX	TYPE "B"(3) MIN	TYPE "B"(3) MAX
1/A	DACK* ACTIVE TO IOR* ACTIVE	DMA	SOURCE	65 [65]		70		70		70	
	DACK* ACTIVE TO IOW* ACTIVE	DMA	DESTIN	310 [305]		310		310		190	
2/A	ADDRESS VALID TO MEMORY COMMAND ACTIVE	DMA	SOURCE DESTIN	91 [89]		88					
3/A	IOR* ACTIVE TO MEMW* ACTIVE	DMA	SOURCE DESTIN	234 [229]							
	MEMR* ACTIVE TO IOW* ACTIVE	DMA	SOURCE DESTIN	0 [0]							
4	DATA FLOAT FROM ACTIVE READ COMMAND	SOURCE	SOURCE		50 [50]		50		50		50
5	DATA VALID FROM IOR* ACTIVE	SOURCE	SOURCE		220 [215]		280		280		160
5/B		SOURCE	DESTIN		242 [237]						
6/A	READ COMMAND ACTIVE HOLD FROM WRITE COMMAND INACTIVE	DMA	SOURCE DESTIN	39 [39]							
7/A	LA<17-23>, SA<0-19>, & SBHE* VALID HOLD FROM MEMORY COMMAND INACTIVE	DMA	SOURCE DESTIN	42 [40]		32					
8	READ DATA VALID HOLD FROM INACTIVE MEMORY COMMAND	SOURCE	SOURCE	0 [0]		2		2		2	
9	IOCHRDY INACTIVE FROM ACTIVE MEMORY COMMAND	SOURCE DESTIN	SOURCE DESTIN								
	8 BITS				384 [366]		398				
	16 BITS				81 [76]		86				
9/B	8 BITS	SOURCE DESTIN	DMA		406 [388]						
	16 BITS				103 [98]						

TABLE 7-1: TRANSFER CYCLE TIMINGS

REF# /DET	DESCRIPTION OF EVENT	DRIVEN BY	MEASURED AT	ISA MIN	ISA MAX [8.33]	COMP. (3) MIN	COMP. (3) MAX	TYPE "A"(3) MIN	TYPE "A"(3) MAX	TYPE "B"(3) MIN	TYPE "B"(3) MAX
9	IOCHRDY ACTIVE FROM ACTIVE MEMORY COMMAND 8 BITS 16 BITS	SOURCE DESTIN	SOURCE DESTIN		373 [355] 70 [65]		398 86				
9/C	8 BITS 16 BITS	SOURCE DESTIN	DMA		462 [444] 159 [154]						
10	TC ACTIVE SETUP TO COMMAND INACTIVE			(4)		(4)		(4)		(4)	
11	TC ACTIVE HOLD FROM COMMAND INACTIVE			(4)		(4)		(4)		(4)	
12/A	IOR* ACTIVE TO INACTIVE (NOTE 2)	DMA	SOURCE	800 [769]		755		395		275	
13/A	IOW* ACTIVE TO INACTIVE (NOTE 2) MEMW* ACTIVE TO INACTIVE (NOTE 2)	DMA DMA	DESTIN DESTIN	489 [469] 489 [469]							
14/A	DACK* ACTIVE HOLD FROM IOR* (NOTE 2) DACK* ACTIVE HOLD FROM IOW* (NOTE 2)	DMA DMA	SOURCE DESTIN	114 [109] 176 [169]		100 155		100 155		35 100	

TABLE 7-1: TRANSFER CYCLE TIMINGS (CONTINUED)

REF# /DET	DESCRIPTION OF EVENT	DRIVEN BY	MEASURED AT	ISA MIN [8.33]	ISA MAX [8.33]	COMP. (3) MIN	COMP. (3) MAX	TYPE "A"(3) MIN	TYPE "A"(3) MAX	TYPE "B"(3) MIN	TYPE "B"(3) MAX
15/A	AEN & BALE VALID FROM I/O COMMAND INACTIVE	PLAT	SOURCE DESTIN	30 [30]		30		30		30	
16	DRQ* ACTIVE HOLD FROM IOR* ACTIVE (NOTE 2)			(5)		(5)		(5)			
16	DRQ* ACTIVE HOLD FROM IOW* ACTIVE (NOTE 2)			(5)		(5)		(5)			
17	IOCHRDY INACTIVE PULSE	SOURCE DESTIN	SOURCE DESTIN	125 [120]	1.56 MICROSECONDS						
18/A	AEN VALID TO I/O COMMAND ACTIVE	PLAT	SOURCE DESTIN	100 [100]		100		100		100	
19	VALID MEMORY READ DATA FROM IOCHRDY ACTIVE	SOURCE	SOURCE								
	8 BITS EVEN				74 [69]		70				
	ODD				64 [59]		70				
	16 BITS				74 [69]		80				
19/A	8 BITS	SOURCE	DESTIN								
	EVEN				85 [80]						
	ODD				75 [70]						
	16 BITS				85 [80]						

TABLE 7-1: TRANSFER CYCLE TIMINGS (CONTINUED)

REF# /DET	DESCRIPTION OF EVENT	DRIVEN BY	MEASURED AT	ISA MIN [8.33]	ISA MAX [8.33]	COMP. (3) MIN	COMP. (3) MAX	TYPE "A"(3) MIN	TYPE "A"(3) MAX	TYPE "B"(3) MIN	TYPE "B"(3) MAX
20/A	IOW* INACTIVE TO ACTIVE (NOTE 2)	DMA	DESTIN	489 [469]		455		335		215	
21/A	IOR* INACTIVE TO ACTIVE (NOTE 2)	DMA	SOURCE	176 [169]		165		165		50	
22/E	COMMAND ACTIVE HOLD FROM IOCHRDY ACTIVE	DMA	SOURCE DESTIN	125 [120]		20		20		20	
23/E	WRITE DATA VALID HOLD FROM INACTIVE I/O COMMAND	SOURCE	DESTIN	25 [25]		20		20		20	
24	VALID READ DATA FROM MEMORY COMMAND ACTIVE (FOR STANDARD CYCLE) 8 BITS 16 BITS	SOURCE	SOURCE		482 [462 173 [165]	490 194					
24/B	VALID READ DATA FROM MEMORY COMMAND ACTIVE (FOR STANDARD CYCLE) 8 BITS 16 BITS	SOURCE	DESTIN		504 [484] 195 [177]						
25	DATA VALID TO IOW* INACTIVE (NOTES 1 & 2)	SOURCE	SOURCE	164 [159]		240		240		130	
25/B		SOURCE	DESTIN	142 [137]							

TABLE 7-1: TRANSFER CYCLE TIMINGS (CONTINUED)

NOTES for TABLE 7-1:

1. The use of 482 [462] in parameter 24 (and the associated 504 [484] in parameter 22/B) will not allow parameter 25 to be met for 8 bit transfers. (This is also parameter 10 in Table 6-1.) If 310 is used for the 8.33 Mhz as used by EISA, then all of the parameters are met. Thus, proposed replacement numbers for parameter 24 (8 bit) are 325 [310] and parameter 24/B (8 bit) 347 [332].

2. These command lines are driven by the Intel 8237A-5.

3. According to the EISA bus specification with the BCLK signal line at 8.33 Mhz.

4. See the TC signal line section of Chapter 5.

5. See the DRQ signal line section of Chapter 5.

SPECIAL DMA CONSIDERATIONS

Each channel of the DMA controller can be programmed for four modes of DMA operation: Single, Block, Demand, or Cascade.

The Cascade mode is used for channels that need to support an add-on bus owner card. Refer to Chapter 9 for more detailed information.

The Single mode allows only a single transfer cycle for each time the DACK* signal line becomes active. Once the DACK* signal line is active, the DRQ signal line can become inactive. If the DRQ signal line remains active, only a single transfer cycle occurs. The add-on slave card requesting service must wait until the DACK* signal line becomes active again before the next transfer cycle.

The Block mode allows sequential block transfer cycles. The transfer cycles continue until the TC (Terminal Count) signal line is driven active by the DMA controller (ISA or E-ISA) or by the add-on slave card (E-ISA).

> *On an ISA platform, a long Block transfer cannot be "interrupted", and refresh cycles may be missed. On an E-ISA platform, the DMA controller will "interrupt" the transfer by driving the DACK* signal line inactive. When it wants to continue the transfer, it re-drives the DACK* signal line active. (The DRQ signal line does not have to be re-driven active.) This arrangement allows the refresh controller and other bus owners access to the bus.*

The Demand mode is similar to the Block mode, except it allows the sequential block transfer to be broken into smaller parts. The sequential block transfer continues until the DRQ signal line is driven inactive. At this time, the present single transfer is completed. When additional parts of the block are ready, the DRQ signal line is driven active. The transfer continues once the DACK* signal line is driven active.

As with the Block mode, the Demand mode is completed when the TC (Terminal Count) signal line is activated by the DMA controller (ISA or E-ISA) or by the add-on slave card (E-ISA).

> *On an ISA platform, if the "smaller parts" of a Demand mode transfer are themselves too large, refresh cycles will be missed as with the Block transfer mode. On an E-ISA platform, the DMA controller can "interrupt" the transfer by driving the DACK* signal line inactive. See the notes in the previous shaded box for more information.*

SPECIAL DRQ AND TC SIGNAL LINE CONSIDERATIONS

The points within a cycle when the DRQ and TC signal lines must be active or inactive varies depending on platform and mode. See the respective sections in Chapter 5 for more information.

STANDARD TRANSFER CYCLE (ISA PLATFORM)

The standard version of the COMPATIBLE cycle is executed only by the platform DMA controller. The data transfer can be from memory to I/O (see Figure 7-1) or from I/O to memory (see Figure 7-2). The resource providing the data is defined as the "source". The resource receiving the data is defined as the "destination".

The request for a transfer cycle must be made by the add-on I/O slave card activating one of the DRQ signal lines. At some unspecified time later, the DMA controller will acknowledge the beginning of a series of transfer cycles with the I/O resource by activating the associated DACK* signal line. (See Figures 7-1 and 7-2.) Before the series of transfer cycles can begin, the AEN and BALE signal lines must be driven active by platform circuitry. An active AEN signal line indicates to all other I/O resources that the forthcoming activity on the ADDRESS and COMMAND signal lines must be ignored. Only the I/O resource that receives the active DACK* signal line may participate in the transfer and respond to active COMMAND signal lines. The memory resource that participates in the transfer is identified by a valid value on the ADDRESS signal lines. An active BALE signal line allows the input circuitry of the add-on memory slave card to receive the valid address. (See the BALE signal line definition in Chapter 5 for more information.)

After the AEN, BALE, and DACK* have been driven active, the series of transfer cycles can begin. To begin each standard transfer cycle, the SA<0-19>, LA<17-23>, and SBHE* signal lines are driven by the DMA controller. (See Figures 7-1 and 7-2.) Programming of the DMA controller determines the transfer direction and data width for the cycle. As outlined in Chapter 4, specific DACK* lines are defined for 8 bit I/O resources, and others are specific for 16 bit resources. The activation of the memory COMMAND signal lines by the DMA controller indicates the direction of the transfer, and that the ADDRESS signal lines are valid. The focus of a transfer cycle is the I/O resource. When the transfer is from the memory to the I/O resource, the data must be valid relative to the deactivation of the IOW* signal line. The memory resource must match this timing. When the transfer is from the I/O resource to the memory resource, the data must be valid within a specific time of activating the IOR* signal line. The memory resource must match this timing.

> *The memory resource in an 8 bit slot must monitor the SMEMR* and SMEMW* signal lines. If either line is active, only the first 1 megabyte of the memory address space is being accessed. For a memory resource in an 8/16 bit slot, the LA<20-23> signal lines are available to indicate the first 1 megabyte. For a memory resource in an 8 bit slot, the LA<20-23>, MEMR*, and MEMW* signal lines are not available. The SMEMR* and SMEMW* signal lines are the only memory command signal lines available in an 8 bit slot; consequently, these memory resources can only reside in the first 1 megabyte of the address space.*

The transfer cycle is completed when both the memory and the I/O COMMAND signal lines are driven inactive. The only requirement for back to back bus cycles is the minimum time between inactive and active I/O COMMAND signal lines.

The length of the standard ISA COMPATIBLE transfer cycle is independent of the width of the data transferred. As previously mentioned, it is focused on the I/O slave resource; consequently, the standard transfer cycle can be referenced to the I/O resource. (See Figures 7-3 and 7-4.) Table 7-2 outlines the types of standard transfer cycles supported for different platforms. The minimum cycle length, in terms of TCLK (period of BCLK) is also summarized. Due to bus settling times and device delays, TCLK should be considered as approximate for the "CYCLE BEGINS" column.

PLAT.	CYCLE TYPE	CYCLE BEGINS	CYCLE ENDS	CYCLE LENGTH	ADD-ON SLAVE CARD
ISA	COMPAT.	IOR* ACT. OR	1 TCLK AFTER IOR* INACT. OR	8 TCLK	MEM, I/O
E-ISA	COMPAT.	2 TCLK TO IOW* ACT.	2 TCLK AFTER IOW* INACT.	8 TCLK	MEM, I/O
E-ISA"	TYPE "A"	IOR* ACT. OR	1 TCLK AFTER IOR* INACT. OR	6 TCLK	I/O
E-ISA"	TYPE "B"	1 TCLK TO IOW* ACT.	2 TCLK AFTER IOW* INACT.(1)	4 TCLK	I/O

TABLE 7-2: ISA & E-ISA STANDARD TRANSFER CYCLES

NOTES:

(1) For Type "B", IOR inactive for 1 TCLK after IOW* inactive.*

STANDARD TRANSFER CYCLE (E-ISA PLATFORM)

Add-on memory E-ISA slave cards involved in the DMA transfer cycle on an E-ISA platform can only operate with COMPATIBLE transfer cycles. The E-ISA platform references IOCHRDY and DRQ signal lines to the BCLK signal line. As with the access cycles, establishing a relationship to the BCLK signal line potentially improves performance. The E-ISA platform has also expanded the DMA transfer cycle definition for add-on I/O slave cards to increase performance. (See Figures 7-5 and 7-6.) The add-on I/O slave card can operate with two additional transfer cycles: TYPE "A", and TYPE "B". These standard transfer cycles differ in standard cycle length. Table 7-2 is a summary of the three types of transfer cycles on an E-ISA platform.

An E-ISA compatible platform also allows each individual DMA channel to be programmed for 8, 16, or 32 bit transfers. When ISA add-on I/O slave cards are involved, only 8 and 16 bit data transfers are supported.

> *The EISA bus specification also defines a BURST transfer cycle. This transfer cycle is only executed between EISA(32) compatible platform resources, and will therefore not be discussed in this book.*

As Table 7-2 indicates, the COMPATIBLE cycle can be executed by either an ISA or an E-ISA platform. Any ISA compatible add-on memory and I/O slave card can participate in a COMPATIBLE cycle. A TYPE "A" or "B" transfer has shorter active I/O COMMAND signal line pulse widths, less data setup time to an active IOW* signal line, and the IOW* signal line becomes active earlier in the cycle. (See Figure 7-7.) All other parameters are the same as a COMPATIBLE cycle.

The reason the EISA bus specification introduces TYPE "A" and TYPE "B" cycles is that ISA compatible add-on I/O slave cards can transfer data faster than COMPATIBLE cycles can transfer data. The concepts are: all ISA compatible add-on I/O slave cards can support COMPATIBLE cycles, most can support a TYPE "A" cycle, and some can support a TYPE "B" cycle. E-ISA compatible add-on I/O slave cards can execute COMPATIBLE, TYPE "A', or TYPE "B" transfers, as outlined in their respective configuration spaces.

Only the ISA COMPATIBLE transfer cycle allows the direct transfer between ISA compatible add-on I/O slave and memory slave cards. For TYPE "A" and TYPE "B" it is assumed that E-ISA platform memory is the memory resource.

> *Figure 7-7 shows the relative positions of the ADDRESS, COMMAND, and DATA signals for the different transfer cycle types. The major differences between the types are shorter COMMAND signal pulse widths and the commensurate shorter valid DATA signals.*

READY TRANSFER CYCLE (ISA PLATFORM)

The ready COMPATIBLE transfer cycle begins in the same fashion as the standard transfer cycle. (See Figures 7-1 and 7-2.) The standard COMPATIBLE transfer cycle becomes a ready transfer cycle if the IOCHRDY signal line is driven inactive by the memory resource. The IOCHRDY signal line must be driven inactive within a "specific time" of an active memory COMMAND signal line. (See Figures 7-8 and 7-9.)

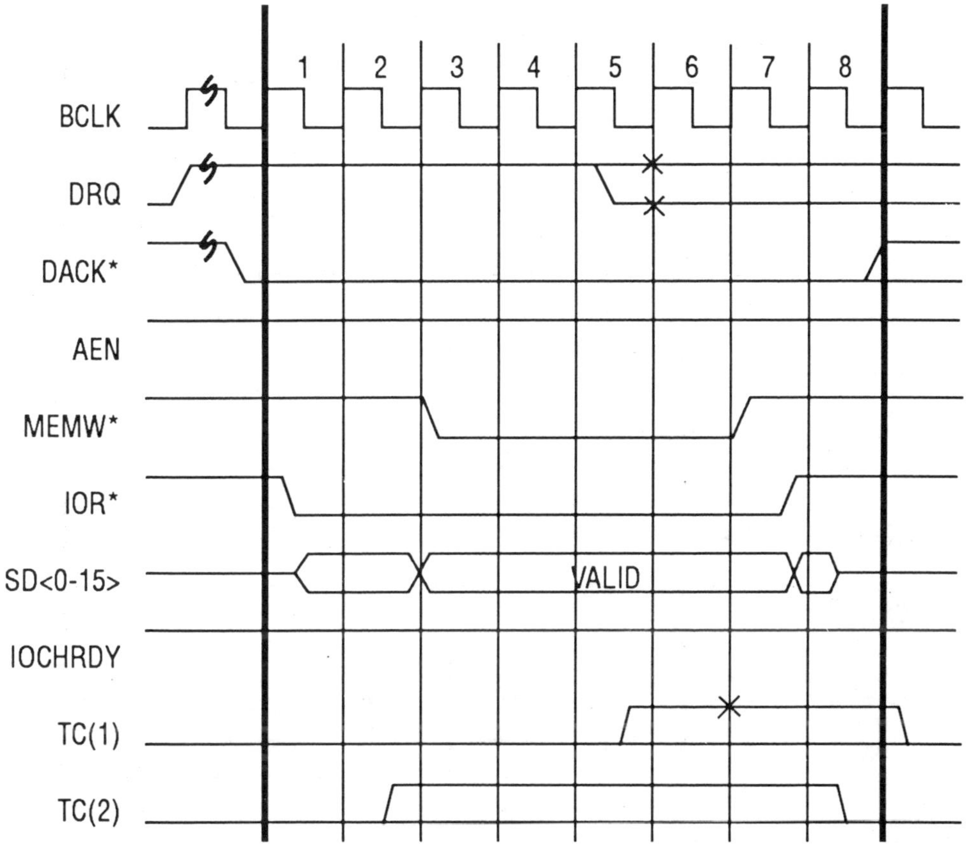

NOTES:
X=Sample Point
(E-ISA only)
(1) Input to Platform
(2) Output from Platform

FIGURE 7-3: STANDARD COMPATIBLE I/O READ TRANSFER CYCLE

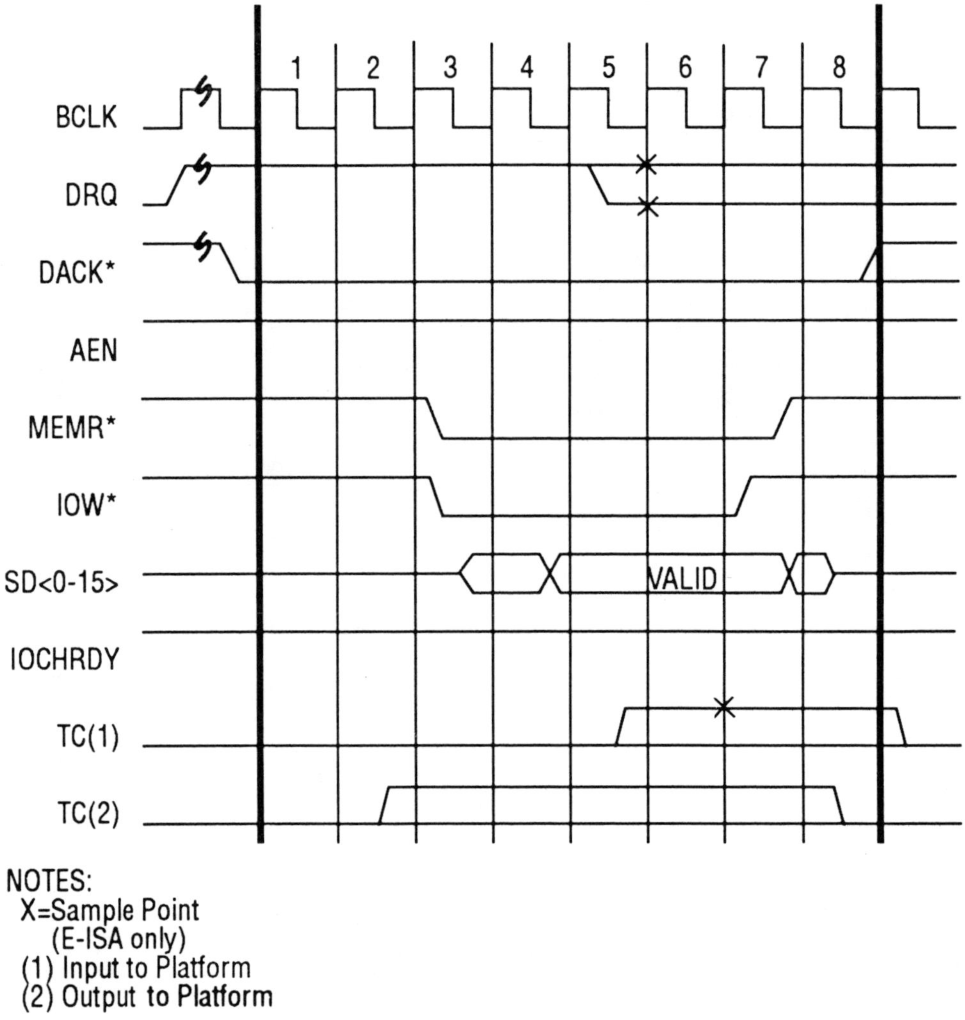

FIGURE 7-4: STANDARD COMPATIBLE I/O WRITE TRANSFER CYCLE

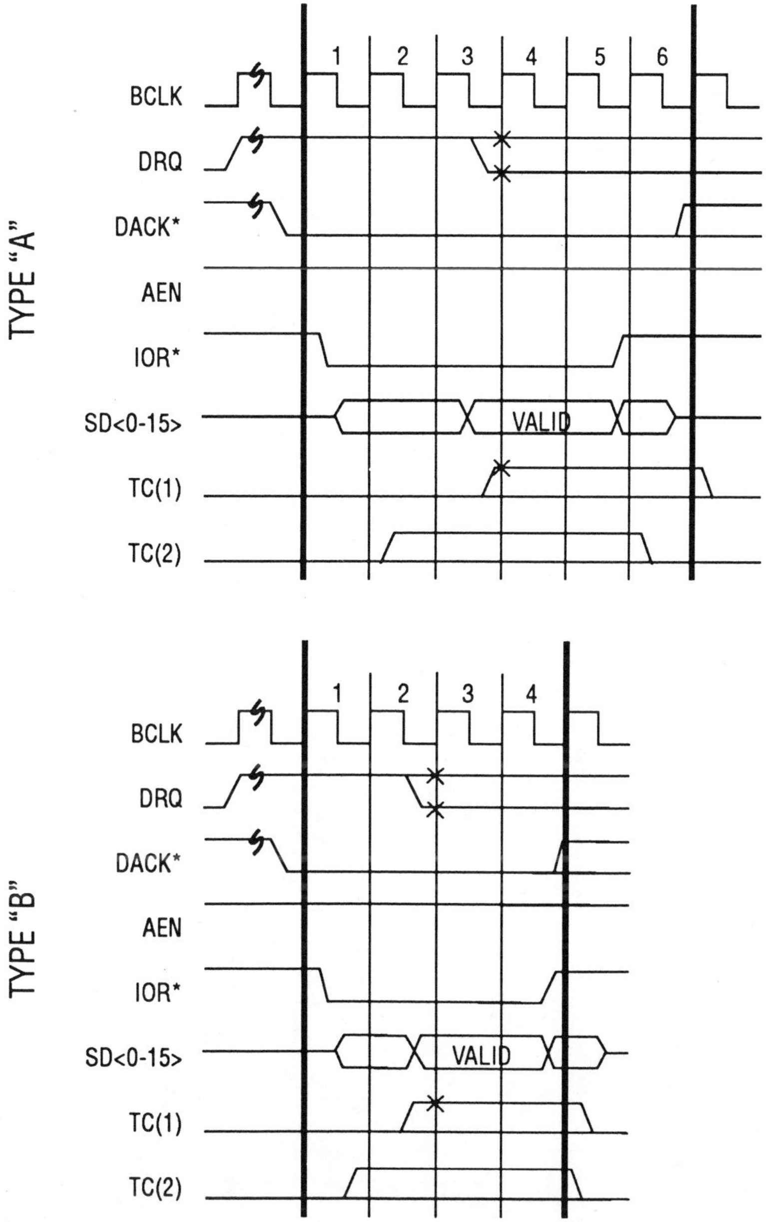

NOTES:
X=Sample Point
(E-ISA Only)
(1) Input to Platform
(2) Output to Platform

FIGURE 7-5: STANDARD E-ISA & EISA-ONLY I/O READ TRANSFER CYCLE

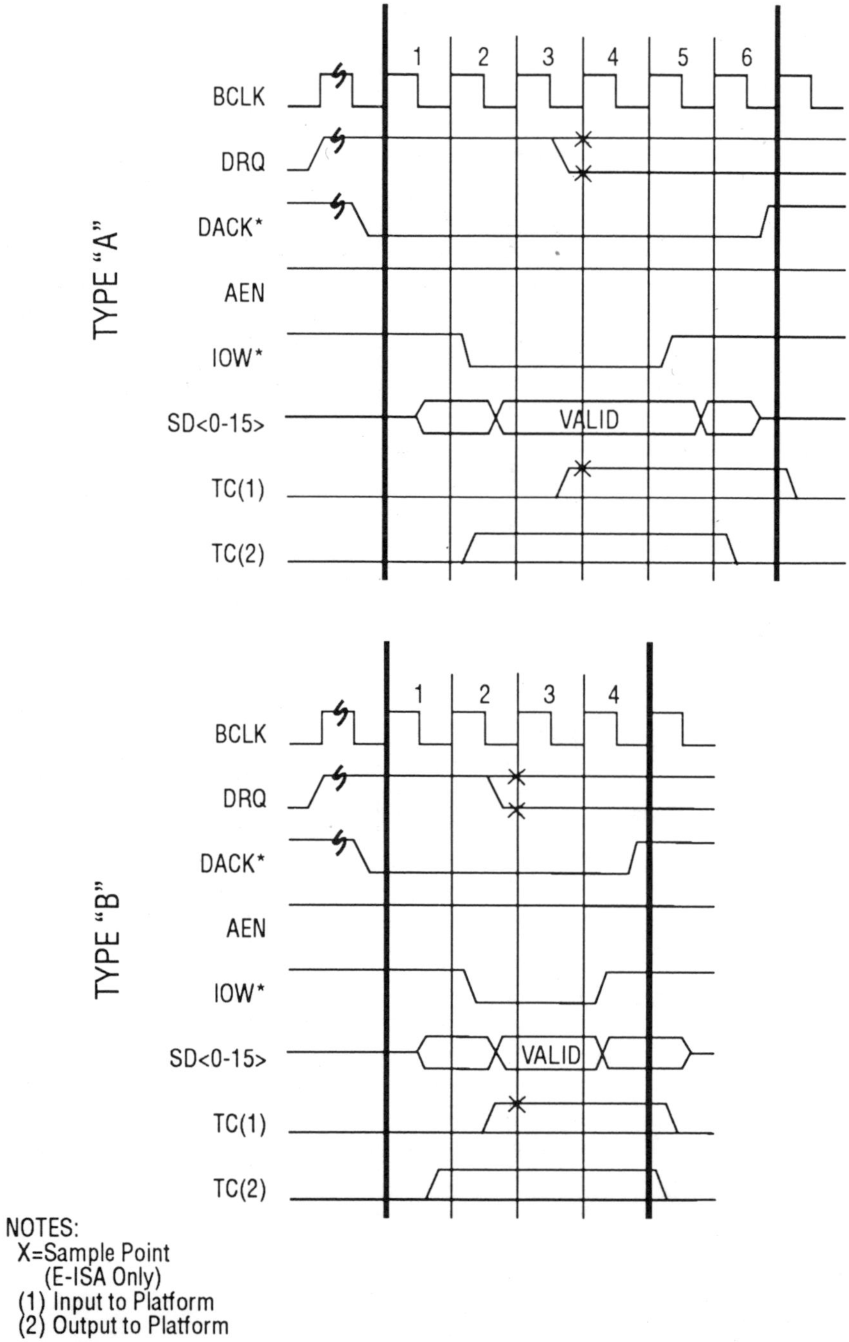

FIGURE 7-6: STANDARD E-ISA & EISA-ONLY I/O WRITE TRANSFER CYCLE

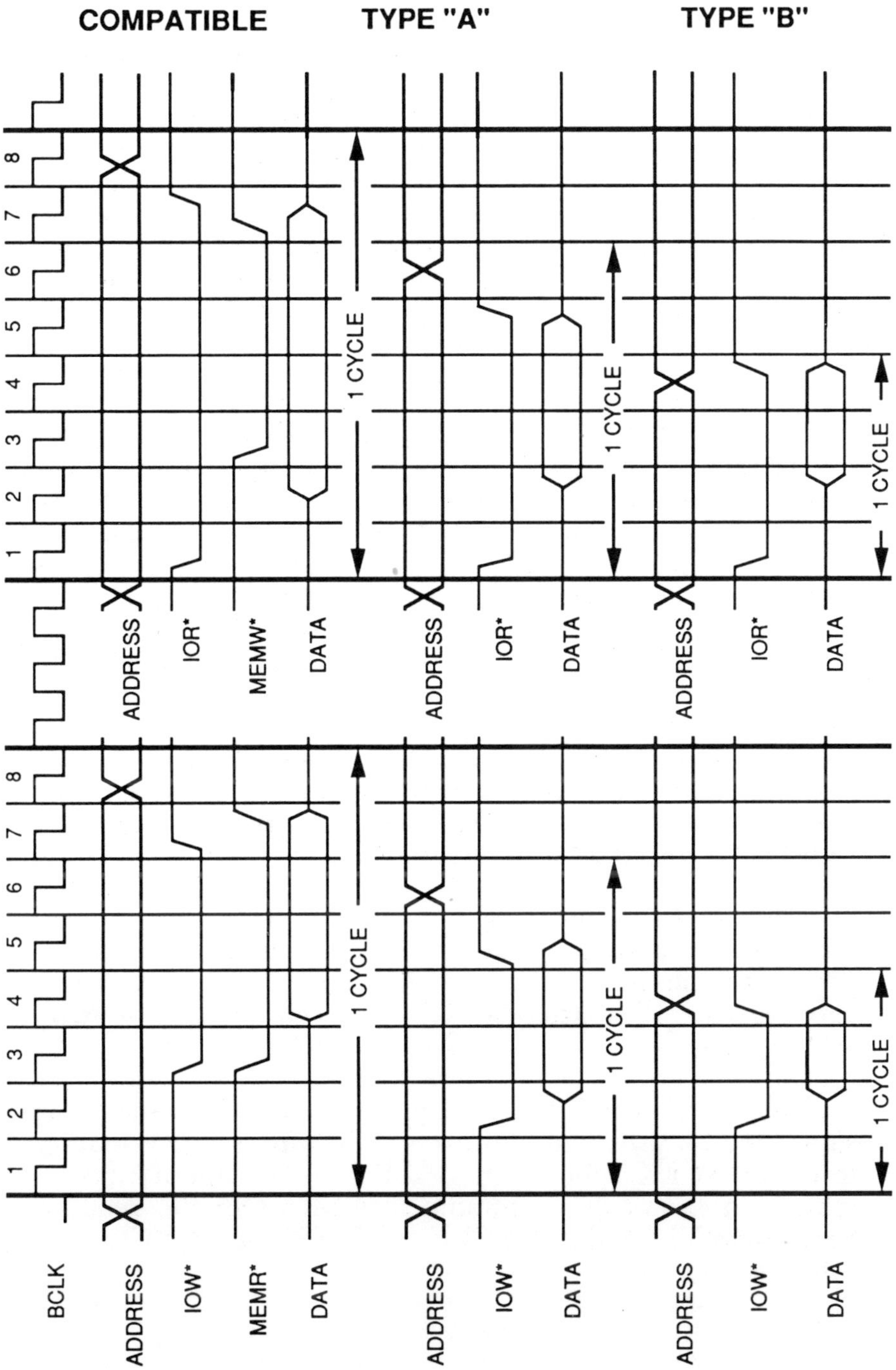

FIGURE 7-7: RELATIVE ADDRESS, COMMAND, & DATA SIGNAL DURATIONS FOR AN E-ISA PLATFORM (STANDARD CYCLE EXAMPLE)

In a transfer cycle, the focal point is the I/O slave resource. However, the transfer requires both a memory resource and an I/O resource. On an ISA platform, the cycle length is controlled only by the memory resource; consequently, data must be valid relative to active I/O COMMAND signal lines. The proper data setup time for a memory write transfer cycle is controlled by the data access time from the I/O resource. It is the responsibility of the memory resource to control the IOCHRDY signal line to insure valid data timing during the transfer. It is the responsibility of the I/O resource to always be able to execute according to standard cycle timings.

The COMMAND signal lines remain active until the IOCHRDY signal line is driven active by the memory resource. During a memory read transfer cycle, the active IOCHRDY signal line indicates to the DMA controller that valid data setup time to an inactive IOW* signal can be met. During a write cycle, it is an indication to the DMA controller that the accessed resource has accepted the data. For either memory read or write transfer cycles, the activation of the IOCHRDY signal line is the indication to the DMA controller to complete the cycle. The transfer cycle is completed when both the memory and the I/O COMMAND signal lines are driven inactive. The only requirement for back to back bus cycles is that the minimum time from inactive to active I/O COMMAND signal lines must be met.

The length of the ready COMPATIBLE transfer cycle is increased in increments of two BCLK signal line periods. As previously mentioned, it is focused on the I/O slave resource; consequently, the basic transfer cycle can be referenced to the I/O resource.

Table 7-3 outlines the types of ready transfer cycles supported for different platforms. The minimum cycle length in terms of TCLK (period of BCLK) is also summarized. The table also shows the minimum increments of the ready cycle. See Table 7-2 for the "CYCLE BEGINS" and "CYCLE ENDS" reference points.

PLAT.	CYCLE TYPE	MIN. CYCLE LENGTH	INCREMENTS OF CYCLE	ADD-ON SLAVE CARD
ISA	COMPATIBLE	10 TCLK	2 TCLK	MEM, I/O
E-ISA	COMPATIBLE	10 TCLK	2 TCLK	MEM, I/O
E-ISA	TYPE "A"	7 TCLK	1 TCLK	I/O
E-ISA	TYPE "B"	5 TCLK	1 TCLK	I/O

TABLE 7-3: ISA & E-ISA READY TRANSFER CYCLES

READY TRANSFER CYCLE (E-ISA PLATFORM)

The operation of the ready compatible transfer cycles by the E-ISA platform is defined in the same way as on an ISA platform. Refer to the READY TRANSFER CYCLE (ISA PLATFORM) section for further information. The only difference between ISA and E-ISA platforms executing a ready transfer cycle is the same as the difference in executing the standard cycle. The E-ISA platform executes two other types of transfer cycles: Type "A" and Type "B". (See Figures 7-10 and 7-11.) Refer to the STANDARD ACCESS CYCLE (E-ISA) section for further information.

As previously discussed, only the ISA COMPATIBLE transfer cycle allows the direct transfer between ISA compatible add-on I/O slave and memory slave cards. For TYPE "A" and TYPE "B" transfers on an E-ISA platform, the memory resource is on the platform.

The various ready transfer cycles have different cycle lengths and increments of cycle lengths. These differences are outlined in Table 7-3.

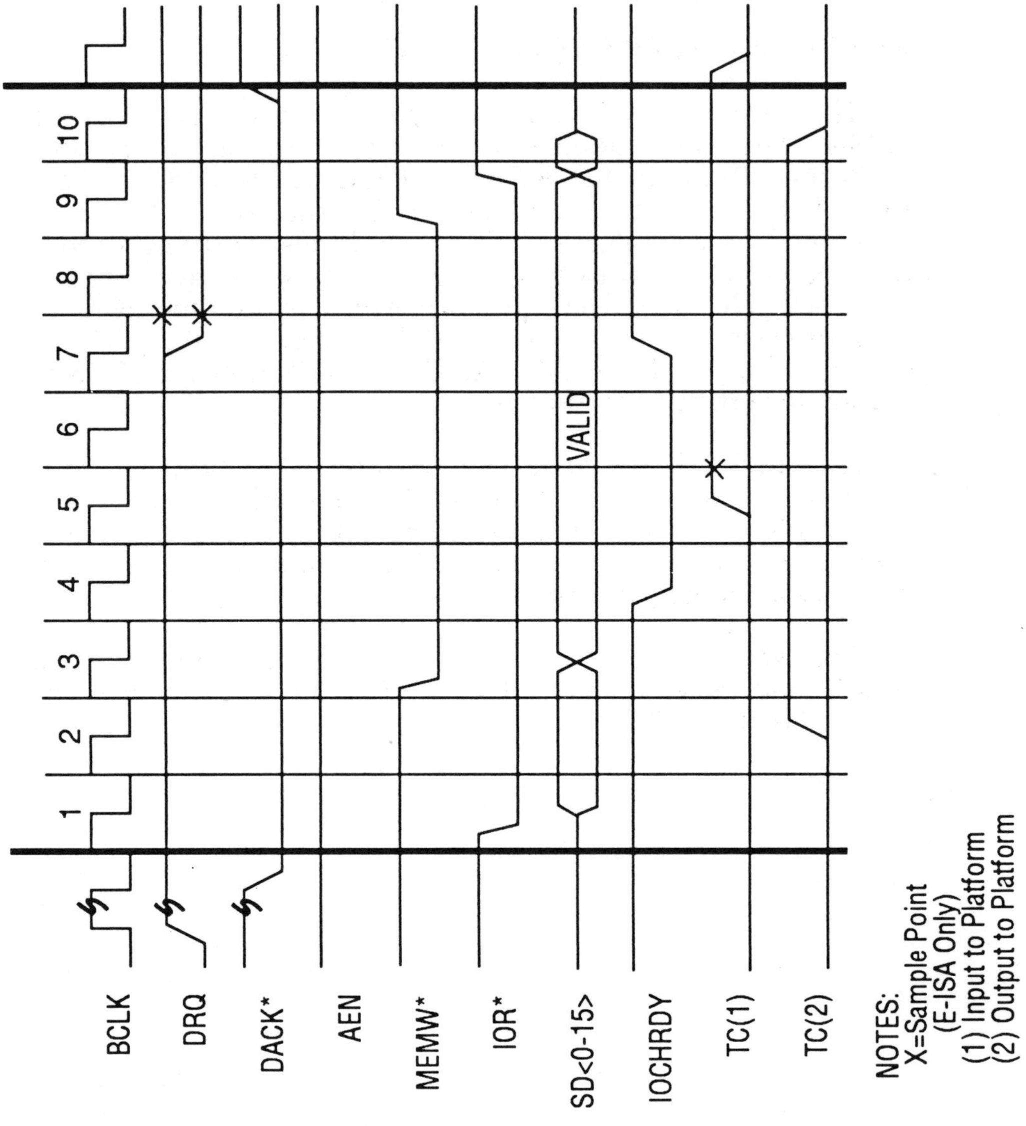

FIGURE 7-8: I/O READ READY TRANSFER CYCLE COMPATIBLE (1 WAIT STATE EXAMPLE)

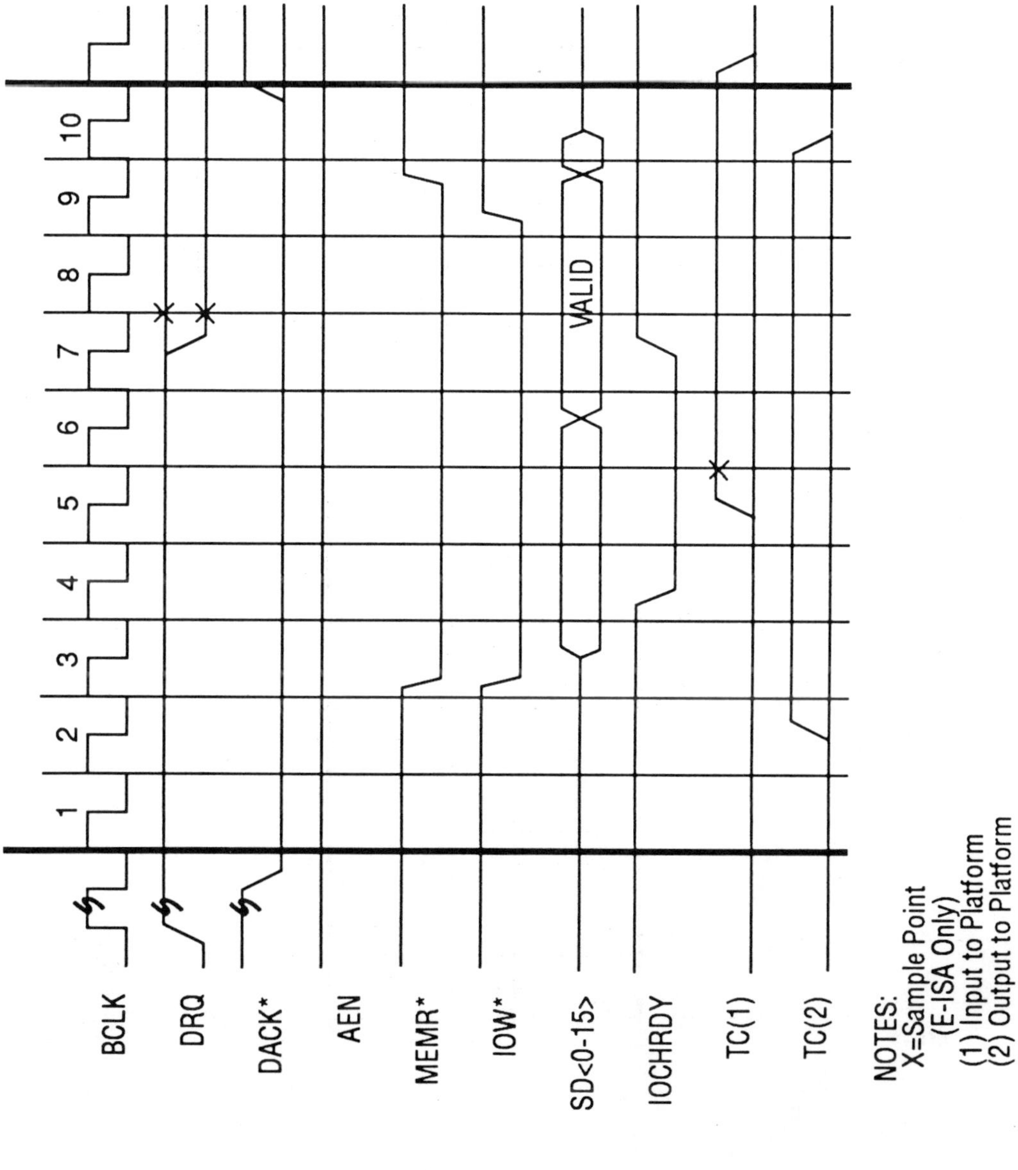

FIGURE 7-9: I/O WRITE READY TRANSFER CYCLE COMPATIBLE (1 WAIT STATE EXAMPLE)

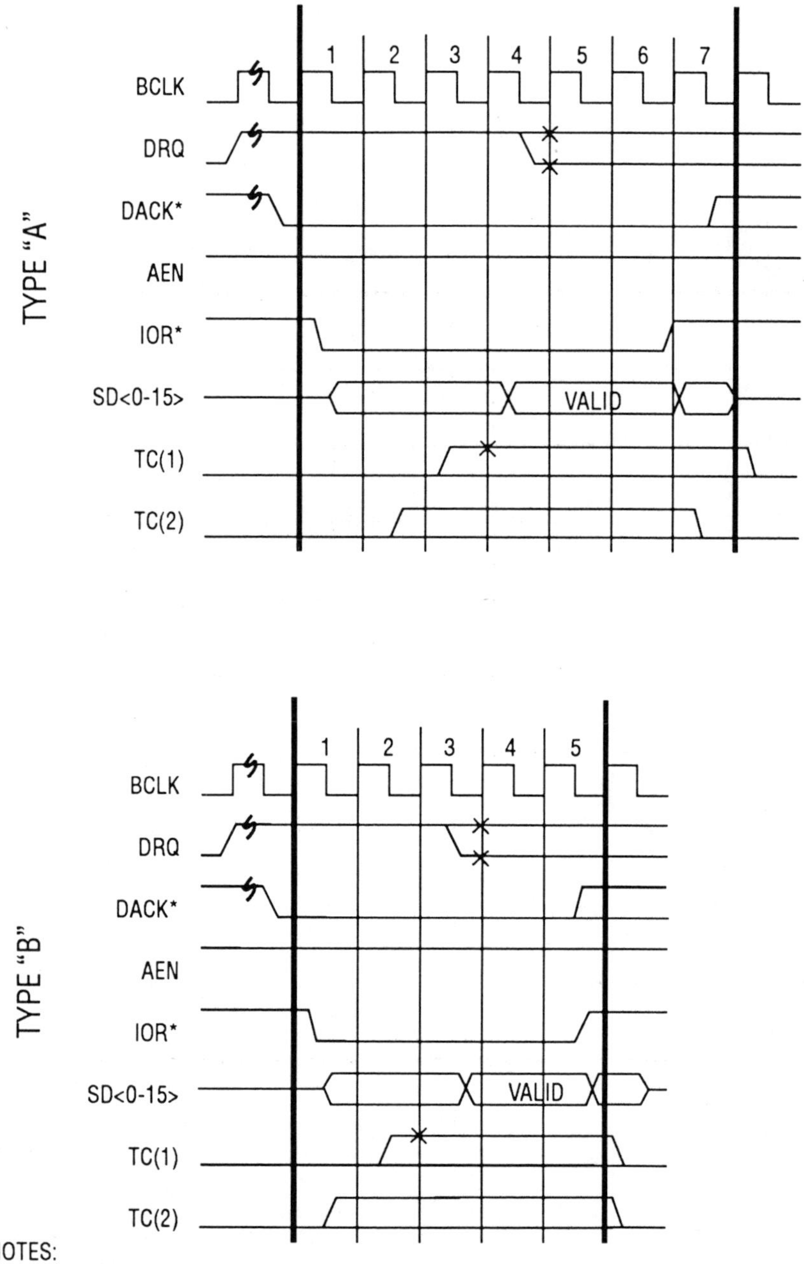

NOTES:
X=Sample Point
(E-ISA Only)
(1) Input to Platform
(2) Output to Platform

FIGURE 7-10: I/O READ READY TRANSFER CYCLE E-ISA & EISA-ONLY (1 WAIT STATE EXAMPLE)

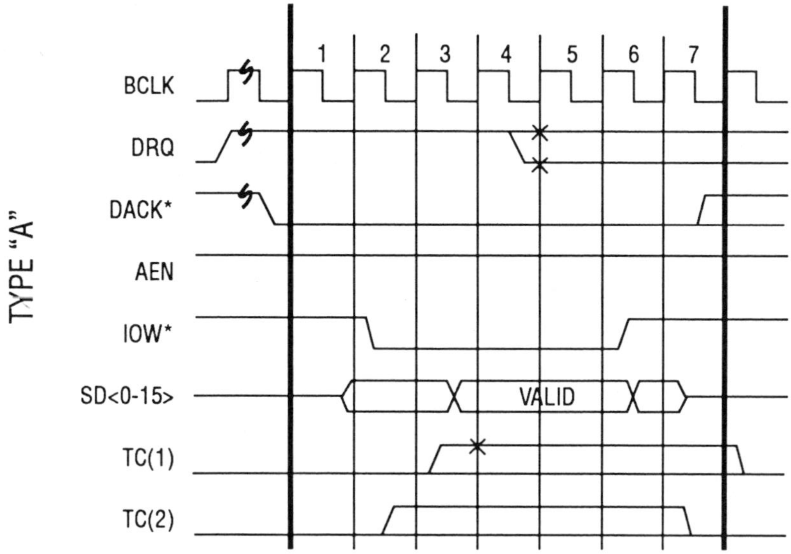

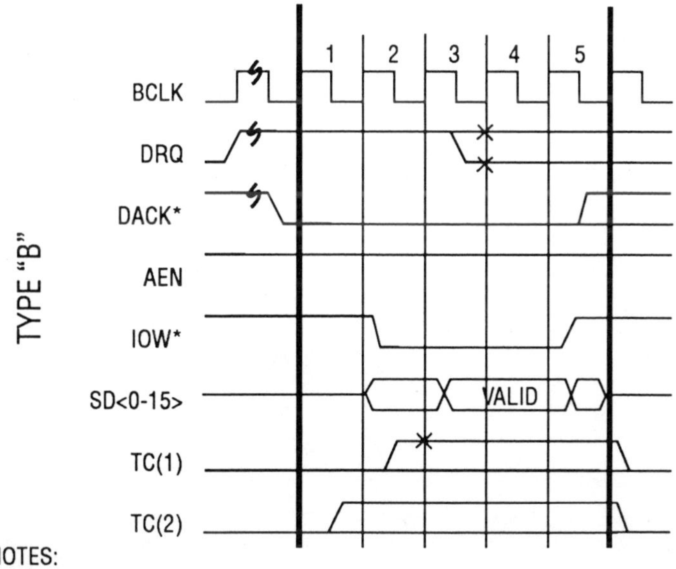

NOTES:
X=Sample Point
(E-ISA Only)
(1) Input to Platform
(2) Output to Platform

FIGURE 7-11: I/O WRITE READY TRANSFER CYCLE E-ISA & EISA-ONLY (1 WAIT STATE EXAMPLE)

CHAPTER EIGHT

REFRESH CYCLES

INTRODUCTION

The refresh cycle is the simplest cycle that is executed on the ISA or E-ISA bus. The purpose of the refresh cycle is to refresh bus DRAM memory resources. During the bus refresh cycle, the platform memory may also be refreshed.

One of the platform timer channels is used to generate a request for a refresh cycle to the platform refresh controller every 15.6 microseconds. If the bus owner is the platform CPU, the refresh cycle takes place immediately. If the bus owner is the DMA controller, the refresh cannot occur until the DMA transfer cycle is complete.

A refresh cycle can also occur when requested by an add-on card that is the current bus owner. The add-on bus owner card obtains bus ownership by "subletting" from the DMA controller; consequently, the refresh will only occur if specifically requested. The add-on bus owner card can request a refresh cycle by simply driving the REF* signal line active.

Table 8-1 outlines the refresh cycle timings for Figure 8-1.

The A.C. timing tables and signal waveform drawings in this section represent a composite of information from several sources in addition to the information that is original to this book. The following sources have provided portions of the A.C. timing and signal waveforms:

"ISA BUS SPECIFICATIONS and APPLICATION NOTES, January 30, 1990", Copyright 1989, Intel Corporation

"TECHNICAL REFERENCE GUIDE, EXTENDED INDUSTRY STANDARD ARCHITECTURE EXPANSION BUS", Copyright COMPAQ COMPUTER Corp., 1989

"PERSONAL COMPUTER BUS STANDARD P996" DRAFT D2.00, January 18, 1990, Copyright IEEE Inc., 1990.

DETAIL	INTERPRETATION (SEE CHAPTER 10 FOR MORE DETAILED INFORMATION)
#	NO SETTLING TIME INCLUDED
#/A	SKEW OF TWO SIGNALS BETWEEN RESOURCES. ONE HAS A DELAY OF 0 NANOSECONDS, AND THE OTHER HAS A DELAY OF 11 NANOSECONDS.
#/B	SIGNAL IS DRIVEN BY ONE RESOURCE TO ANOTHER. THE SECOND RESOURCE MUST RESPOND BY DRIVING A SIGNAL BACK TO THE FIRST RESOURCE. # INCLUDES THE "ROUND TRIP" TIME OF 11 + 11 = 22 NANOSECONDS.
#/C	SAME AS "B", EXCEPT ONE OF THE RESOURCES RELIES ON A 1K OHM PULL-UP RESISTOR. # INCLUDES THE "ROUND TRIP" TIME OF 11 + 78 = 89 NANOSECONDS.
#/D	SAME AS "B", EXCEPT ONE OF THE RESOURCES RELIES ON A 300 OHM PULL UP RESISTOR. # INCLUDES THE "ROUND TRIP" TIME OF 11 + 22 = 33 NANOSECONDS.
#/E	SAME AS "A", EXCEPT THE WORST CASE INTERPRETATION IS THAT BOTH SIGNALS ARE EITHER 0 NANOSECONDS OR 11 NANOSECONDS. THE NET RESULT IS A DIFFERENCE OF 0 NANOSECONDS.

"DRIVEN BY" AND "MEASURED BY" INTERPRETATION

ADD-ON	ADD-ON ACCESSED RESOURCE
ISA-B	ISA OR E-ISA COMPATIBLE ADD-ON CARD BUS OWNER
CPU	PLATFORM CPU
PLAT	PLATFORM CIRCUITRY

MIN. AND MAX. INTERPRETATION

ALL TIMES ARE IN NANOSECONDS WITH AN 8.00 MHZ BCLK.
THE NUMBER IN "[ ]" REFLECTS THE CALCULATED VALUE FOR AN 8.33 MHZ BCLK.

EISA INTERPRETATION

THE ENTRIES ARE THOSE IN THE EISA BUS SPECIFICATION MEASURED AT THE PLATFORM.

SMEMR* AND SMEMW* HAVE A 10 NANOSECOND DELAY FROM MEMR* AND MEMW*.

REF# /DET	DESCRIPTION OF EVENT	DRIVEN BY	MEASURED AT	MIN	MAX	EISA
				[8.33]		
1/A	MEMR* ACTIVE TO INACTIVE	REF.	ADD-ON	214 [204]		235
2/A	ADDRESS VALID SETUP TO MEMR* ACTIVE (NOTE 2)	REF.	ADD-ON	70		70 [70]
3/A	ADDRESS HOLD HOLD TO MEMR* INACTIVE	REF.	ADD-ON	25 [25]		25
4	IOCHRDY INACTIVE FROM MEMR* ACTIVE	ADD-ON	ADD-ON		81 [76]	86
4/B	IOCHRDY INACTIVE FROM MEMR* ACTIVE	ADD-ON	REF.		103 [98]	
4	IOCHRDY ACTIVE FROM MEMR* ACTIVE	ADD-ON	ADD-ON		70 [65]	86
4/B	IOCHRDY ACTIVE FROM MEMR* ACTIVE	ADD-ON	REF.		159 [154]	
5/B	MEMR* INACTIVE FROM IOCHRDY ACTIVE	REF.	ADD-ON	125 [120]		116
6/A	REF* ACTIVE SETUP TO MEMR* ACTIVE	REF.	ADD-ON	114 [109]		120
7/A	REF* ACTIVE HOLD FROM MEMR* INACTIVE	REF.	ADD-ON	20 [20]	239 [229]	20
8/A	ADDRESS & MEMR* TRISTATE FROM MEMR* INACTIVE	REF.	ADD-ON		125 [120]	
9	IOCHRDY ACTIVE TO INACTIVE	ADD-ON	ADD-ON	125 [120]		
10	PLATFORM OWNERSHIP FROM REF* INACTIVE	PLAT.	ADD-ON	250 [240]		
13/A	REF* ACTIVE FROM VALID ADDRESS	PLAT.	ADD-ON	0		
14/A	REF* HOLD FROM VALID ADDRESS	PLAT.	ADD-ON	0		
15	ADD-ON CARD TRISTATE OF ADDRESS AND CONTROL LINES BEFORE REF* ACTIVE	ADD-ON	ADD-ON	0		

TABLE 8-1: REFRESH CYCLES

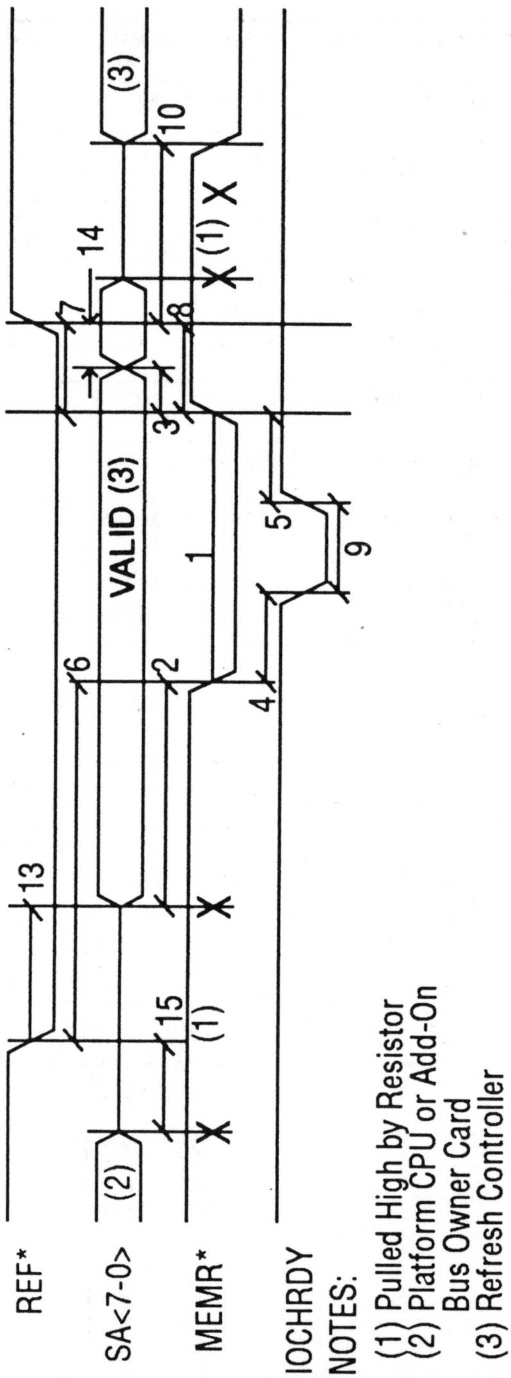

FIGURE 8-1: REFRESH CYCLES

REFRESH CYCLE (ISA PLATFORM)

The resource driving the ADDRESS and MEMR* signal lines must first tristate these lines before the beginning of the refresh cycle. For platform resources, these signal lines must be tristated prior to allowing the refresh controller to become the bus owner. The refresh controller then drives the REF* signal line to begin the cycle. For the add-on bus owner card, these signal lines must be tristated prior to the add-on bus owner card driving the REF* signal line active.

The refresh controller begins the cycle by driving the ADDRESS signal lines with the valid refresh address and subsequently driving the MEMR* signal line active. The cycle will be a standard refresh cycle if the IOCHRDY signal remains active. A standard cycle ends with the refresh controller driving the MEMR* signal line inactive and the ADDRESS signal lines invalid. If the add-on card requested the refresh cycle, the add-on card must drive the REF* signal line inactive to complete the cycle. If an add-on bus owner card did not request the refresh cycle, the refresh controller will drive the REF* signal line inactive to complete the cycle.

> *When the add-on bus owner card requests the refresh cycle, it must drive the REF* signal line inactive within a specific time of the MEMR* signal line being driven inactive. Additional refresh cycles will be executed if the REF* is driven active beyond this specific time.*

If the IOCHRDY signal line is driven inactive within a specific time from the MEMR* signal line, the cycle becomes a ready refresh cycle. The IOCHRDY* signal line can only be driven by the memory resources requiring longer refresh cycles. When the IOCHRDY signal line is driven inactive, the refresh controller will drive the MEMR* signal line inactive to complete the cycle.

After the MEMR* signal line is driven inactive, the resource that relinquished control of the bus for the refresh cycle can again drive the signal lines.

> *The other COMMAND signal lines must not be driven active by any resource during the refresh cycle. Pull-up resistors on the bus backplane insures that these signal lines are pulled up inactive. The other ADDRESS and DATA signal lines should be tristated by all other resources to prevent buffer fights. All the other signal lines, such as MEMCS16*, IOCS16*, AEN, etc. are undefined.*

The cycle length of a refresh cycle is measured from the point of tristating to the point of redriving the ADDRESS and MEMR* signal lines. For a standard refresh cycle, this time is 554 nanoseconds (534 nanoseconds for an 8.33 Mhz BCLK). The increment of cycle length for a ready refresh cycle is a single BCLK period.

REFRESH CYCLE (E-ISA PLATFORM)

The refresh cycle on the E-ISA platform is executed in the same fashion as on the ISA platform, except for two enhancements. First, the IOCHRDY signal lines are sampled relative to the clock edges of the BCLK signal line. Second, the refresh address is driven onto the SA<0-15> signal lines.

CHAPTER NINE

ARBITRATION CYCLES

INTRODUCTION

The ISA and E-ISA platforms allow add-on cards to become the bus owners. The add-on card must execute an arbitration cycle to obtain bus ownership. The arbitration takes place between other add-on cards that want to be bus owners or want to execute DMA transfer cycles. The central control point for the arbitration cycle is the DMA controller.

An ISA compatible add-on card that is bus owner can execute only standard and ready access cycles. An E-ISA compatible add-on bus owner card can also execute no wait state cycles, which reference the SRDY* signal line to the BCLK signal line. See the SPECIAL E-ISA NOTE section in Chapter 1 for more information.

No DMA transfer cycles are supported because the bus is "subleased" from the DMA controller. (See Chapter 4 for more information.) In addition, the add-on bus owner can request a refresh cycle to be executed.

The EISA(32) specification allows add-on cards to become bus owners in two ways. First, it supports the ISA and E-ISA method of "subleasing" from the DMA controller. Second, an add-on bus master may use the discrete bus request and bus acknowledge lines which are available on six of the EISA(32) slots. For purposes of this book, only the method supported on the ISA and E-ISA platforms will be discussed.

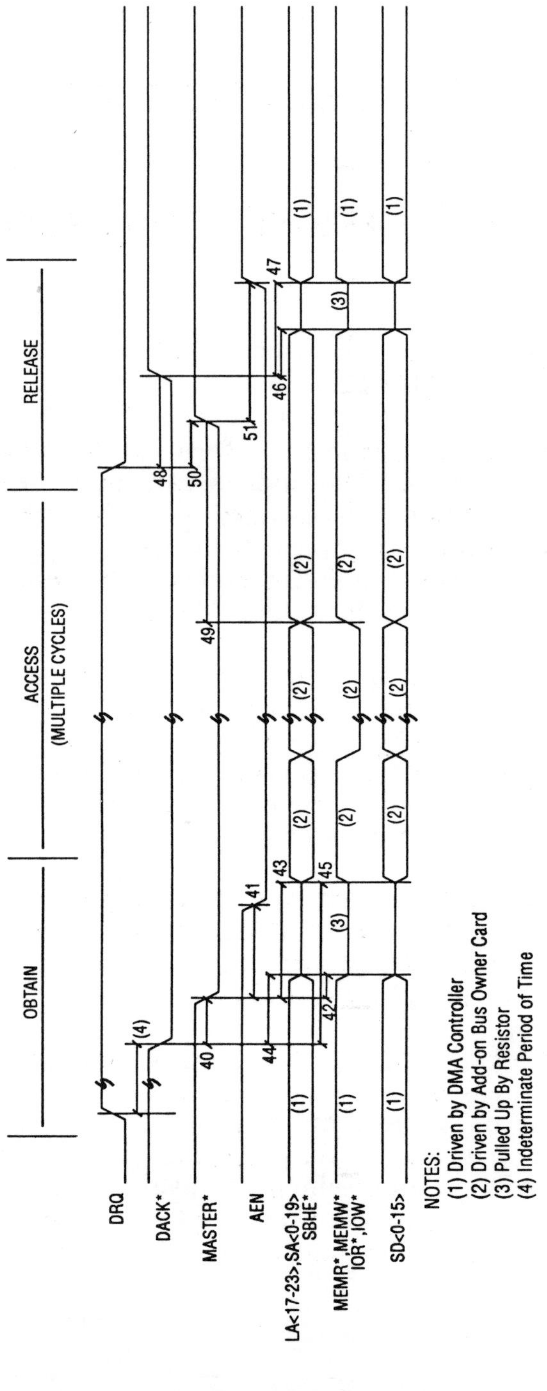

FIGURE 9-1: ARBITRATION CYCLES

Table 9-1 summarizes the timing parameters for Figure 9-1.

Table 9-1 has several key pieces of information. First, the "#/DET" (parameter number / detail) column relates the parameter number of the table to Figure 9-1. Second, it provides detailed information on how the number was derived. The timing numbers are affected by the settling time on the bus. The table is based on the default values for 8 slots as outlined in Chapter 10. The interpretation of the #/DET entries are shown below.

The A.C. timing tables and signal waveform drawings in this section represent a composite of information from several sources in addition to the information that is original to this book. The following sources have provided portions of the A.C. timing and signal waveforms:

"ISA BUS SPECIFICATIONS and APPLICATION NOTES, January 30, 1990", Copyright 1989, Intel Corporation

"TECHNICAL REFERENCE GUIDE, EXTENDED INDUSTRY STANDARD ARCHITECTURE EXPANSION BUS", Copyright COMPAQ COMPUTER Corp., 1989

"PERSONAL COMPUTER BUS STANDARD P996" DRAFT D2.00, January 18, 1990, Copyright IEEE Inc., 1990.

DETAIL	INTERPRETATION
#	NO SETTLING TIME INCLUDED. (SEE CHAPTER 10 FOR MORE DETAILED INFORMATION)
#/A	SKEW OF TWO SIGNALS BETWEEN RESOURCES. ONE HAS A DELAY OF 0 NANOSECONDS, AND THE OTHER HAS A DELAY OF 11 NANOSECONDS.
#/B	SIGNAL IS DRIVEN BY ONE RESOURCE TO ANOTHER. THE SECOND RESOURCE MUST RESPOND BY DRIVING A SIGNAL BACK TO THE FIRST RESOURCE. # INCLUDES THE "ROUND TRIP" TIME OF 11 + 11 = 22 NANOSECONDS.
#/C	SAME AS "B", EXCEPT ONE OF THE RESOURCES RELIES ON A 1K OHM PULL-UP RESISTOR. # INCLUDES THE "ROUND TRIP" TIME OF 11 + 78 = 89 NANOSECONDS.
#/D	SAME AS "B", EXCEPT ONE OF THE RESOURCES RELIES ON A 300 OHM PULL UP RESISTOR. # INCLUDES THE "ROUND TRIP" TIME OF 11 + 22 = 33 NANOSECONDS.
#/E	SAME AS "A", EXCEPT THE WORST CASE INTERPRETATION IS THAT BOTH SIGNALS ARE EITHER 0 NANOSECONDS OR 11 NANOSECONDS. THE NET RESULT IS A DIFFERENCE OF 0 NANOSECONDS.

"DRIVEN BY" AND "MEASURED BY" INTERPRETATION

ADD-ON	ADD-ON ACCESSED RESOURCE
ISA-B	ISA OR E-ISA COMPATIBLE ADD-ON CARD BUS OWNER
CPU	PLATFORM CPU
PLAT	PLATFORM CIRCUITRY

MIN. AND MAX. INTERPRETATION

ALL TIMES ARE IN NANOSECONDS WITH AN 8.00 MHZ BCLK.
THE NUMBER IN "[ ]" REFLECTS THE CALCULATED VALUE FOR AN 8.33 MHZ BCLK.

REF# /DET	DESCRIPTION OF EVENT	DRIVEN BY	MEASURED AT	MIN [8.33]	MAX	EISA
40	MASTER* ACTIVE FROM DACK* ACTIVE	ISA-B	ISA-B	0 [0]		
41/B	AEN INACTIVE FROM MASTER* ACTIVE	PLAT	ADD-ON	0 [0]	71 [71]	
42/A	ADDRESS, COMMAND, & DATA LINES TRISTATED FROM MASTER* ACTIVE (NOTE:1)	PLAT	ISA-B	0 [0]	60 [60]	
43	ISA-B DRIVES ADDRESS, COMMAND, & DATA LINES FROM MASTER* ACTIVE	ISA-B	ISA-B	125 [120]		
44/A	ADDRESS, COMMAND, & DATA LINES TRISTATED FROM DACK* ACTIVE (NOTE:1)	PLAT	ISA-B	0 [0]	60 [60]	60
45	ISA-B DRIVES ADDRESS, COMMAND, & DATA LINES FROM DACK* ACTIVE	ISA-B	ISA-B	125 [120]		120
46	ADDRESS, COMMAND, & DATA LINES TRISTATED FROM DACK* INACTIVE	ISA-B	ISA-B		60 [60]	60
47	ADDRESS, COMMAND, & DATA LINES DRIVEN FROM DACK* INACTIVE	PLAT	ISA-B	60 [60]		
48/A	DACK* INACTIVE FROM DRQ INACTIVE	PLAT	ISA-B	375 [360]		300
49	MASTER* INACTIVE FROM COMMAND INACTIVE	ISA-B	ISA-B	60 [60]		
50	MASTER* INACTIVE FROM DRQ* INACTIVE	ISA-B	ISA-B		100 [100]	
51/B	AEN ACTIVE FROM MASTER* INACTIVE	PLAT	ADD-ON	0 [0]	71 [71]	

TABLE 9-1: ADD-ON BUS OWNER ARBITRATION CYCLES

NOTES:

(1) No settling time is associated with tristate propagation on the bus.

ARBITRATION CYCLES (ISA PLATFORM)

The access cycle portion of the arbitration cycle is executed by an add-on bus owner card in the same fashion as the standard and ready access cycles are executed by an ISA platform CPU. The only additional feature of the cycle is that at the beginning and end of the arbitration cycle there is an obtain and release portion of the cycle, respectively. Figure 9-1 outlines the obtain, standard access, ready access, and release portions of the arbitration cycles.

The obtain portion of the cycle begins when the bus owner card activates the DRQ signal line and waits for an active DACK* signal line. (See Figure 9-1.) In response to the active DACK* request signal line, the add-on bus owner activates the MASTER* signal line. The platform circuitry responds to the active MASTER* signal line by driving the AEN signal line inactive. Simultaneously, the DMA controller tristates the ADDRESS, CONTROL and DATA signal lines. After the MASTER* signal line goes active, the add-on bus owner card drives the ADDRESS, CONTROL, and DATA signal lines.

The release portion of the cycle begins when the add-on bus owner card drives the MASTER* and DRQ signal lines inactive. (See Figure 9-1.) In response to the inactive DRQ signal line, the DMA controller drives the DACK* signal line inactive. The inactive MASTER* signal line causes platform circuitry to drive the AEN signal line inactive. An inactive DACK* signal line causes the add-on bus owner card to tristate the ADDRESS, COMMAND, and DATA signal lines. Also, subsequent to an inactive DACK* signal line, the DMA controller circuitry drives the ADDRESS, COMMAND, and DATA signal lines.

The standard or ready access cycles that are executed by the add-on bus owner card can have the same bus timings as the platform CPU. According to the IEEE P996 specification, the add-on slave cards and other bus resources cannot distinguish the source of the cycles. The EISA bus specification does document a different set of timing numbers for the ISA compatible add-on bus owner card. The EISA bus specification states that if the add-on bus owner card is synchronized to

BCLK, the timings for the platform CPU can be used. If the add-on bus owner card is not synchronized to BCLK, a different set of numbers should be used. This different set of numbers reflects the measurements of existing add-on cards.

ARBITRATION CYCLES (E-ISA PLATFORMS)

The obtain and release portions of the arbitration cycle on an E-ISA platform operate in the same fashion as on an ISA platform.

CHAPTER TEN

BUS INTEGRITY

INTRODUCTION

The use of a bus to interconnect add-on cards to platform resources generates a number of unique design requirements. These requirements are caused by effects that range from the "transmission line" nature of the bus traces to the effects of the bus on the input and output structures of the add-on boards.

The IBM technical reference manuals do not completely specify the backplane characteristics of trace separation, impedance, inductance, and so forth; consequently, bus backplanes developed by clone manufacturers have varied widely. Also, the IBM technical reference manuals simply state that the load of each slot should equal two low-power Schottky loads.

The IEEE P996 and the EISA bus specifications have attempted to more uniformly define the backplane. In addition, Intel has developed a high reliability AT compatible backplane which defines the backplane characteristics, DIN connector, and signal line placement on a EUROcard form factor. See Chapter 14 for more detailed information.

The diversity of the backplanes and add-on cards, the various interpretations of the bus, and the lack of engineering sophistication by some platform and board vendors requires a very conservative approach to the various topics in this chapter.

D.C. LOADING AND A.C. TIMING CONSIDERATIONS

Whenever the logic level of a signal line changes, there is a transition period before the signal line has settled to a new stable logic level at all the slot locations on the bus. The settling time is dependent on the length of the backplane, the total number of loaded slots, the load of each slot, the construction of the backplane, whether termination devices are used to minimize signal reflections, and so forth. In addition, the length of a signal line from the connector to the driver or receiver, and the types of drivers or receivers on the add-on cards and platforms vary greatly. The IEEE P996 committee, the EISA bus specification, and Intel have attempted to place "REASONABLE" parameters on slot interface to maximize compatibility and reliability. A summary of these parameters for ISA and E-ISA slot interfaces is shown in Table 10-1.

> *The EISA bus specification configuration space allows 15 logical slots in addition to the single (Slot 0) logical platform. However, all of the EISA timing specifications are based on seven add-on card slots in addition to the platform.*

MAXIMUM LENGTH FROM DRIVER OR RECEIVER TO THE CONNECTOR	2.5 INCHES MAX. (3)
SPACING BETWEEN SLOT CONNECTORS	.8 INCHES MAX.
CAPACITIVE LOAD OF PLATFORM & PER SLOT	15 pF. (1)
V_{ih} AT PLATFORM OR SLOT V_{il} AT PLATFORM OR SLOT	2.0 VOLTS MIN. .8 VOLTS MAX.
V_{oh} AT PLATFORM OR SLOT V_{ol} AT PLATFORM OR SLOT	2.4 VOLTS MIN. .5 VOLTS MAX.
I_{ih} AT PLATFORM OR PER SLOT I_{il} AT PLATFORM OR PER SLOT	40 uA MAX. (2) -400 uA MAX. (2)
I_{oh} AT PLATFORM OR PER SLOT I_{ol} AT PLATFORM OR PER SLOT	SEE TABLES 10-2 & 10-3

TABLE 10-1: ISA & E-ISA PLATFORM & SLOT INTERFACE

NOTES:

(1) EISA allows 20 pF per slot.
(2) The EISA bus specification allows twice the stated current. The values in the table reflect the values from the original IBM technical reference manuals, which specify two LS loads per slot.
(3) The EISA bus specification specifies a 7.5 inch maximum distance from connector to platform driver or receiver.

The output drive characteristics of the platform and add-on cards have also been defined beyond the original IBM technical reference manuals. Some busses only specify one type of bus driver for all signal lines. On ISA and E-ISA platforms there are several types of drivers.

ISA and E-ISA platforms and add-on cards use low-power Schottky (LS) or advanced low-power Schottky (ALS) driver devices to limit the rise and fall times of the signal lines. This approach minimizes the settling time of the signals. The EISA bus specification does not state any such requirement.

The ISA and E-ISA busses require three types of drivers: tristate (TS), totem pole (TP), and open collector (OC). The TS drivers are used on signal lines that are bi-directional or can be driven by different resources at different times. The TP drivers are used on signal lines that are only driven by one resource all the time. The OC drivers are used on signal lines that are driven by different resources at different times or several resources at the same time.

Each type of driver can have two minimum current requirements: version "1" and version "2". The existence of two versions of each driver is partly historical. The original IBM platforms had "line driver" quality drive devices for some signal lines, while other signal lines simply were outputs from "normal" TTL logic devices. The other reason for the existence of two versions is that some signal lines only have one slot load. For example, on ISA platforms, the TC signal line is driven only by the DMA controller, and only seven of the 15 slots can have DMA channels.

Table 10-2 is a summary of the output current requirements of the different drivers.

DRIVER TYPE	DRIVER VERSION	I_{ol} MIN. mA ISA	I_{ol} MIN. mA E-ISA	I_{oh} MIN. mA ISA	I_{oh} MIN. mA E-ISA
TP	1	2	5	-1	-.4
	2	24	24	-3	-3
TS	1	2	5	-1	-.4
	2	24	24	-3	-3
OC	1	8	5	-	-
	2	20	24	-	-

TABLE 10-2: OUTPUT CURRENT REQUIREMENTS

Table 10-3 is a summary of the drive devices' pull-up resistors for each signal line. The pull-up resistors only reside on the platform, not on the add-on cards.

LINE	DRIVER TYPE (1)	PULL UP (3) ISA	PULL UP (3) E-ISA	LINE	DRIVER TYPE	PULL UP (3) ISA	PULL UP (3) E-ISA
AEN	TP2[TP1]	-	-	MEMR*	TS2[TS2]	4.7K	8.2K
BALE	TP2[TP2]	-	-	MEMW*	TS2[TS2]	4.7K	8.2K
DACK*	TP1[TP1]	-	-	OSC	TP2[TP2]	-	-
DRQ	TP2[TS1]	-	(2)	REF*	OC1[OC2]	4.7K	300
IOCHCK*	OC1[OC1]	4.7K	4.7K	RESET	TP2[TP2]	-	-
IOCHRDY	OC1[OC2]	1.0K	1.0K	SA<0-19>	TS2[TS2]	-	-
IOCS16*	OC1[OC2]	300	300	SBHE*	TS2[TS2]	-	-
IOR*	TS2[TS2]	4.7K	8.2K	SD<0-7>	TS2[TS2]	-	8.2K
IOW*	TS2[TS2]	4.7K	8.2K	SD<8-15>	TS2[TS2]	-	8.2K
IRQ	TP2[OC1]	-	8.2K	SMR/SMW*	TP2[TP2]	-	-
LA<17-23>	TS2[TS2]	-	-	SYSCLK	TP2[TP2]	-	-
MASTER*	OC1[OC2]	300	300	TC	TP1[TS2]	-	-
MEMCS16*	OC1[OC2]	300	300	SRDY*	OC1[OC2]	300	300

TABLE 10-3: OUTPUT DRIVER TYPES

NOTES:

(1) The bracketed "[ ]" numbers are for E-ISA compatible platform and add-on cards

(2) Pull-down resistor of 5.6K

(3) See the IEEE P996 specification for the exact resistor value and location for the PC/96 high-reliability bus from Intel, or contact the Intel Multibus II group concerning the P2/aPC bus.

There has never been a specification for AT bus signal line impedance or termination. In addition, very few platforms have any terminations at all. The IEEE P996 has recommended that for increased "data integrity and system reliability" termination networks be added that match the characteristic impedance of the backplane. It recommends that a 40-60 ohm resistor and a 30-70 pF capacitor be installed at either end of the signal lines as shown below:

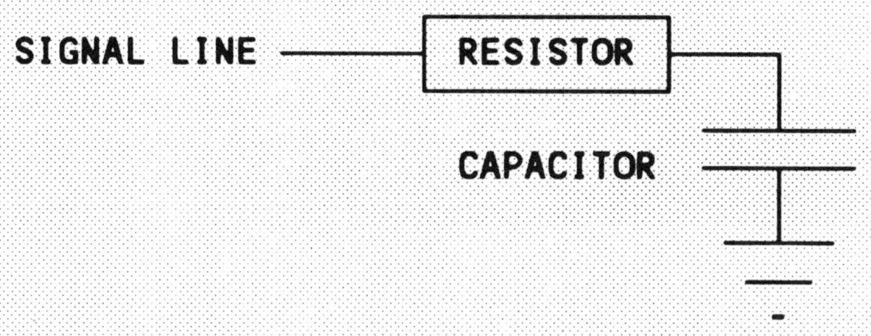

A.C. timings are one of the most important considerations for reliable bus operations. Because IBM and other manufacturers have never published a comprehensive set of A.C. timing specifications, designers have had to "back engineer" from IBM platforms. The situation has now improved, with Intel (for the P2/aPC connector on Multibus II), IEEE (through the P996 technical committee), and COMPAQ (as part of the EISA consortium) providing A.C. timings for the bus. Intel and the IEEE have adopted identical A.C. timing parameters. The numbers provided by COMPAQ (the EISA bus specification) are in excellent agreement. Any differences reflect differing interpretations of bus settling times and the use of a higher BCLK signal line frequency (8.33 Mhz versus 8.00 Mhz).

In addition to the lack of timing specifications, problems are compounded further because many platforms and add-on cards have been developed without proper bus engineering. Trace spacing and width, layer location, and impedance have varied widely. The trace length from the connector to the drivers and receivers sometimes far exceeds 2.5 inches. Furthermore, the type and number of drivers and receivers have varied widely.

In order to resolve these problems while maintaining compatibility with existing platforms and add-on cards, the IEEE P996 committee, Intel, and this author have assumed an 11 nanosecond bus settling time. The EISA bus specification has left it to the add-on card designer to establish the settling time.

The 11 nanosecond bus settling time allowance is more than sufficient to account for backplane impedance, trace length violations, load violations, slot location of driver and receiver, and most other similar conditions. It is based on the assumption of eight slots being used. Use of additional slots will require the use of a longer settling time.

The inclusion of 11 nanoseconds within the timing parameters will insure a very high degree of compatibility between past, present, and future platforms and boards. It assumes a fully loaded 8 slot backplane (the platform is the ninth slot). For a low to high transition, the time is measured from .5 volts at the driver to 2.0 volts at the receiver. For a high to low transition, the time is measured from 2.4 volts at the driver to .8 volts at the receiver.

This book has adopted the 11 nanosecond settling time in its A.C. timing tables. In that 11 nanoseconds can be considered an "over kill" for properly designed backplanes and add-on cards, all of the timing tables include information as to how the 11 nanosecond and associated timing delays can be interpreted. A designer can then adjust these numbers, if necessary, to reflect the actual implementation.

The interpretation of the "#/DET" column in the A.C. timing tables of this book is as follows:

- #: The timing number with no Detail indicates that no bus settling was included. An example is parameter 5 of Table 6-1. A valid LA<17-23> signal line value is measured at the add-on card. A decoding of the address occurs on the add-on card and the MEMCS16* signal line is properly driven.

Because both are measured at the add-on card, no bus settling time is included.

- **#/A**: These timing numbers include the skew between two signals driven by the same source and monitored by another. An example is parameter 4/A of Table 6-1. The platform CPU drives the LA<17-23> signal lines; subsequently, it drives the memory COMMAND signal line active. The platform CPU (for 16 bits) has a minimum of 120 nanoseconds between these two events. When measured at the add-on card, it is listed as 109 nanoseconds. This reflects the worst case assumption that the ADDRESS signal lines used the full 11 nanoseconds settling time while the COMMAND signal line has the absolute minimum settling time of 0 nanoseconds. This represents the theoretical worst case; the actual worst case for a reasonably designed or lightly loaded system is less.

- **#/B**: These timing numbers reflect bus settling in two directions. A "driving" resource drives a signal line to a "receiving" resource on the bus. The receiving resource responds by driving a signal back to the "driving" source. An example is parameter 5/B of Table 6-1. The platform CPU "drives" valid the LA<17-23> signal lines to the add-on card, and it takes 11 nanoseconds for the signal to settle at the add-on card. The add-on card decodes the ADDRESS signal lines and responds by driving the MEMCS16* signal line active. The platform CPU will not receive a stable MEMCS16* signal line until 11 nanoseconds after it was driven active by the add-on card. Thus, in addition to the 90 nanosecond decode time (see parameter 5), there is an additional 22 nanoseconds of "round trip" settling time before the MEMCS16* signal line is valid relative to the LA<17-23> signal lines at the platform CPU.

- **#/C** and **#/D**: These parameters are similar to the #/B parameter in that a "round trip" settling time is involved. In the case of the #/B parameter, the "round trip" time was 11 + 11 = 22 nanoseconds. In the case of the #/C parameter, the signal line is "driven" active only by a 1K ohm pull-up resistor. Thus, the "round trip" is 11 + 78 = 89 nanoseconds. In the case of the #/D parameter, the signal line is "driven" active by a 300 ohm pull-up resistor. Thus, the "round trip" is 11 + 22 = 33 nanoseconds.

An example of the #/C parameter is 20/C of Table 6-1. The platform CPU will not receive a stable IOCHRDY signal line until 78 nanoseconds after the open collector driver of the add-on card has "released" it. Thus, in addition to the 373 nanoseconds (see parameter 20 for 16 bits), there is an additional 89 nanoseconds of "round trip" settling time before the IOCHRDY signal line is valid relative to an active COMMAND signal line at the platform CPU.

An example of the #/D parameter is 5/D of Table 6-1. The platform CPU will not receive a stable MEMCS16* signal line until 22 nanoseconds after the open collector driver of the add-on card has "released" it. Thus, in addition to the 66 nanoseconds (see parameter 5 for MEMCS16* signal line inactive), there is an additional 33 nanoseconds of "round trip" settling time before the MEMCS16* signal line is valid relative to valid LA<17-23> signal lines at the platform CPU.

- **#/E**: This parameter is the same as the #/A parameter, except that the theoretical worst case settling time is 0 nanoseconds. An example of this parameter is 6/E of Table 6-1. When the platform CPU drives the LA<17-23> signal lines invalid there is a minimum hold time before the add-on card responds by driving the MEMCS16* signal line invalid. In this case, a pull-up resistor will drive the MEMCS16* signal line inactive. But from the view point of worst case hold time, 0 nanoseconds is used. The actual minimum hold time will be greater.

CHAPTER ELEVEN

PERFORMANCE

Tables 11-1 to 11-7 summarize the maximum bus performance for back to back access and transfer cycles. These tables are based on the BCLK signal being 8.00 Mhz for the ISA bus and 8.33 Mhz for the E-ISA bus. The transfer cycles do not include any overhead for the time delay between active DRQ and DACK* signal lines.

DATA SIZE (BITS)	ISA [E-ISA] ACCESS (MEGABYTES PER SECOND)		
	NO WAIT	STANDARD	READY
8	NA [2.77]	1.33 [1.39]	1.19 [1.19]
16	8.00 [8.33]	5.33 [5.55]	4.00 [4.165]

TABLE 11-1: ISA & E-ISA PLATFORM CPU ACCESS MEMORY PERFORMANCE

DATA WIDTH (BITS)	ISA [E-ISA] ACCESS (MEGABYTES PER SECOND)		
	NO WAIT	STANDARD	READY
8	NA [2.77]	1.33 [1.39]	1.19 [1.19]
16	NA [NA]	5.33 [5.55]	4.00 [4.165]

TABLE 11-2: ISA & E-ISA PLATFORM CPU ACCESS I/O PERFORMANCE

DATA SIZE (BITS)	ISA [E-ISA] ACCESS (MEGABYTES PER SECOND)	
	STANDARD	READY
8	1.33 [1.39]	1.19 [1.19]
16	5.33 [5.55]	4.00 [4.165]

TABLE 11-3: ISA & E-ISA ADD-ON BUS OWNER CARD ACCESS MEMORY PERFORMANCE

DATA SIZE (BITS)	ISA [E-ISA] ACCESS (MEGABYTES PER SECOND)	
	STANDARD	READY
8	1.33 [1.39]	1.19 [1.19]
16	5.33 [5.55]	4.00 [4.165]

TABLE 11-4: ISA & E-ISA ADD-ON BUS OWNER CARD ACCESS I/O PERFORMANCE

DATA SIZE (BITS)	ISA [E-ISA] TRANSFER (MEGABYTES PER SECOND)	
	STANDARD	READY
8	1.00 [1.04]	.80 [.83]
16	2.00 [2.08]	1.60 [1.67]

TABLE 11-5: ISA & E-ISA PLATFORM COMPATIBLE TRANSFER PERFORMANCE

DATA SIZE (BITS)	[E-ISA] TYPE "A" TRANSFER (MEGABYTES PER SECOND)	
	STANDARD	READY
8	[1.39]	[1.19]
16	[2.78]	[2.38]

TABLE 11-6: E-ISA PLATFORM TYPE "A" TRANSFER PERFORMANCE

DATA SIZE (BITS)	[E-ISA] TYPE "B" TRANSFER (MEGABYTES PER SECOND)	
	STANDARD	READY
8	[2.08]	[1.67]
16	[4.17]	[3.33]

TABLE 11-7: E-ISA PLATFORM TYPE "B" TRANSFER PERFORMANCE

CHAPTER TWELVE

CONFIGURATION SPACE

INTRODUCTION

As mentioned in Chapter 2, there are specific I/O address locations assigned to each of the add-on card slots on the EISA bus. The AEN signal lines are specific to each slot. An access to an I/O address reserved for a specific slot results in the deactivation of the AEN signal line for that slot. The other AEN signal lines are activated to prevent the other add-on slave I/O cards from responding to the access cycle.

RESERVED REGISTERS

Within each set of the specific I/O address locations assigned to a specific slot there are five locations defined by the EISA bus specification. These registers are at addresses 0XC80H to 0XC84H. The value of X is the slot number with the value of X=0 reserved for the platform.

The information contained in registers 0XC80H to 0XC83H consists of platform and add-on card manufacturers' product and revision codes. Address 0XC80H also contains a not ready indicator.

The register at address location 0XC84H for the add-on cards contains control bits. The register allows the add-on cards to be individually enabled and reset. The register also has a bit that can be set by the add-on card to indicate that it was the source of an active IOCHCK signal line.

These configuration registers may be supported on an E-ISA platform, but are not supported on an ISA platform. The platform CPU will read an FFH from the 0XC80H location if the add-on card does not support this function or if no add-on card is present.

See the EISA bus specification for further information concerning the use of these registers.

The EISA bus specification suggests that FFH is first written to 0XC80H before the register is read. As outlined by the EISA bus specification, this write cycle "precharges" the bits contained in this register. On a platform that supports this feature, the EISA bus specification identifies an 8.2K pull up resistor on the data lines. The DC load of each slot is specified by the IEEE P996 to be 40 uA; consequently, the 8.2K provides a 2.6 volt level on the data lines for an 8 slot backplane (including one "slot" used by the platform). The pull-up resistors on the data lines of platforms with more than 8 slots must have smaller pull-up resistors to insure at least a 2.0 voltage level.

<u>*CHAPTER THIRTEEN*</u>

POWER AND INITIALIZATION

POWER TOLERANCE

The power to add-on cards is supplied only though the slot connectors, and must meet specific voltage and current requirements. Table 13-1 outlines these requirements.

VOLTAGE NOMINAL	VOLTAGE MIN	VOLTAGE MAX	CURRENT 8 MAX	CURRENT 8/16 MAX	VOLTAGE SENSE MIN	PEAK/PEAK NOISE MAX (5)
12	11.4	12.6	1.5 (1)	1.5 (1)	10.8	120 mV
-12	-10.8 (2)	-13.2 (2)	.3	.3	-10.2	120 mV
5	4.875 (3)	5.25	3.0	4.5	4.5	50 mV
-5	-4.5 (4)	-5.5 (4)	.2	.2	-4.3	50 mV

TABLE 13-1: POWER PER SLOT

NOTES:

(1) IEEE P996 specs 1.0
(2) IEEE P996 specs -11.4 and -12.6
(3) EISA specs 4.5
(4) IEEE P996 specs -4.75 and -5.25
(5) PEAK to PEAK noise voltage is defined by the IEEE P996

The above table represents the combination of the EISA and the IEEE P996 bus specifications. For the cases in which the IEEE P996 bus specification disagrees with the EISA bus

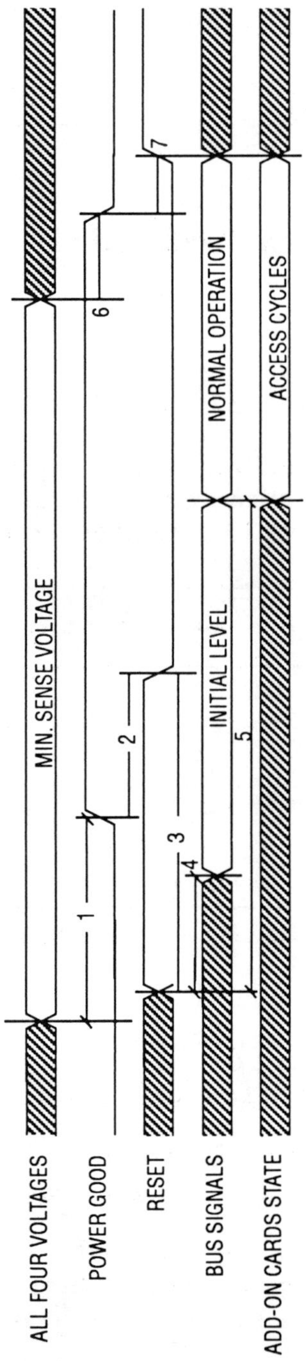

FIGURE 13-1: POWER ON SEQUENCE

specification, the entry placed in the table for voltage is the one that was closest to the original IBM PC, XT and AT Technical Reference Manuals.

The minimum 5 volt signal of 4.875 is specified by the IEEE P996 and is higher than the 4.5 specified by the EISA bus specification and the IBM Technical Reference Manual. The higher voltage allows for the fact that some chips require a minimum of 4.5 volts.

INITIALIZATION

The VOLTAGE SENSE entries in Table 13-1 are from the original IBM PC and XT Technical Reference Manuals. The power supply in an IBM, ISA, E-ISA, or EISA system drives a power good signal to the reset control circuitry of the platform CPU. This signal remains inactive for a period between 100 and 500 milliseconds (#1) after the four voltages are above the minimum SENSE VOLTAGE. (See Figure 13-1.) The reset control circuitry on the platform drives the RESET signal line active during this period. The RESET signal is held active a minimum of 1 millisecond (#2) after the power good signal is active or until the RESET signal line is active for a minimum of nine BCLK signal line periods (TCLK), whichever is longer (#3). When the RESET signal line is driven active, the add-on cards have a maximum of 500 nanoseconds (#4) to drive the bus signal lines to their initial state. Also, there is a maximum of 1 millisecond (#5) from an active RESET signal line before the platform CPU can begin access cycles to the add-on cards.

Whenever any of the voltages drop below the SENSE VOLTAGE, the power good signal from the power supply will be driven inactive immediately (#6). (See Figure 13-1.) In response to an inactive power good signal, the RESET signal line is immediately driven active (#7).

The initial state of all signal lines is inactive except for the BCLK, OSC, IRQ, ADDRESS, and DATA signal lines. The BCLK and OSC signal lines are driven at their appropriate frequency. The ADDRESS and DATA signal lines are driven by the platform CPU but are invalid. The IRQ signal lines are active unless an add-on card has specifically driven individual lines inactive.

> *Right after reset, the platform CPU is the first bus owner.*

> *All of the above information concerning initialization is a combination of information from the IEEE P996 specification and various IBM technical reference manuals. It is sometimes unclear which platforms adhere to these specifications.*

CHAPTER FOURTEEN

MECHANICAL NOTES

The following pages show some of the physical characteristics of two versions of the AT bus. The first version is the original and most widely used connector scheme, and the second is the new high reliability version, proposed by Intel and appended to the IEEE P996 specification.

The first four figures show board outlines, bracket position, and connector pin identification for the original PC/AT bus. The connectors are readily available in either one- or two-piece construction. They are typically mounted on 0.8 inch centers on a platform CPU board or on a passive backplane.

The high reliability version is implemented on a EUROcard form factor, and uses a 96 pin DIN connector. It is referred to as the PC/96 bus. The PC/96 bus improves the ground/signal line ratio, provides a larger card form factor, and has a more reliable connector for use in hostile environments. The connector has three rows of 32 pins each, and has the male part of the connector on the bus. This arrangement allows the optional use of a 128 pin connector on the card, leaving one row of pins available for user defined functions.

The connector used for the PC/96 bus is the DIN 41632. The EUROcard form factor is specified in the mechanical portion of ANSI/IEEE Std 1296-1987. This specification is an application of ANSI/IEEE Std 1101-1987.

See the IEEE P996 specification for additional information on the PC/96 bus. You may also obtain information from the Intel Multibus II group by asking about the P2/aPC bus.

	Signal	Pin	Pin	Signal
	0V	B01	A01	IOCHCHK*
	RESET	B02	A02	SD7
	+5V	B03	A03	SD6
	IRQ9	B04	A04	SD5
	-5	B05	A05	SD4
	DRQ2	B06	A06	SD3
	-12V	B07	A07	SD2
	SRDY*	B08	A08	SD1
	+12V	B09	A09	SD0
	0V	B10	A10	IOCHRDY
	SMEMW*	B11	A11	AEN
	SMEMR*	B12	A12	SA19
	IOW*	B13	A13	SA18
	IOR*	B14	A14	SA17
P1	DACK3*	B15	A15	SA16
	DRQ3	B16	A16	SA15
	DACK1*	B17	A17	SA14
	DRQ1	B18	A18	SA13
	REFRESH*	B19	A19	SA12
	SYSCLK	B20	A20	SA11
	IRQ7	B21	A21	SA10
	IRQ6	B22	A22	SA9
	IRQ5	B23	A23	SA8
	IRQ4	B24	A24	SA7
	IRQ3	B25	A25	SA6
	DACK2*	B26	A26	SA5
	TC	B27	A27	SA4
	BALE	B28	A28	SA3
	+5V	B29	A29	SA2
	OSC	B30	A30	SA1
	0V	B31	A31	SA0

Bracket End ↑

← Add-in Board Component Side

	Signal	Pin	Pin	Signal
	MEMCS16*	D01	C01	SBHE*
	IOCS16*	D02	C02	LA23
	IRQ10	D03	C03	LA22
	IRQ11	D04	C04	LA21
	IRQ12	D05	C05	LA20
	IRQ15	D06	C06	LA19
	IRQ14	D07	C07	LA18
	DACK0*	D08	C08	LA17
	DRQ0	D09	C09	MEMR*
P2	DACK5*	D10	C10	MEMW*
	DRQ5	D11	C11	SD8
	DACK6*	D12	C12	SD9
	DRQ6	D13	C13	SD10
	DACK7*	D14	C14	SD11
	DRQ7	D15	C15	SD12
	+5V	D16	C16	SD13
	MASTER*	D17	C17	SD14
	0V	D18	C18	SD15

FIGURE 14-1: ISA BUS PIN IDENTIFICATION

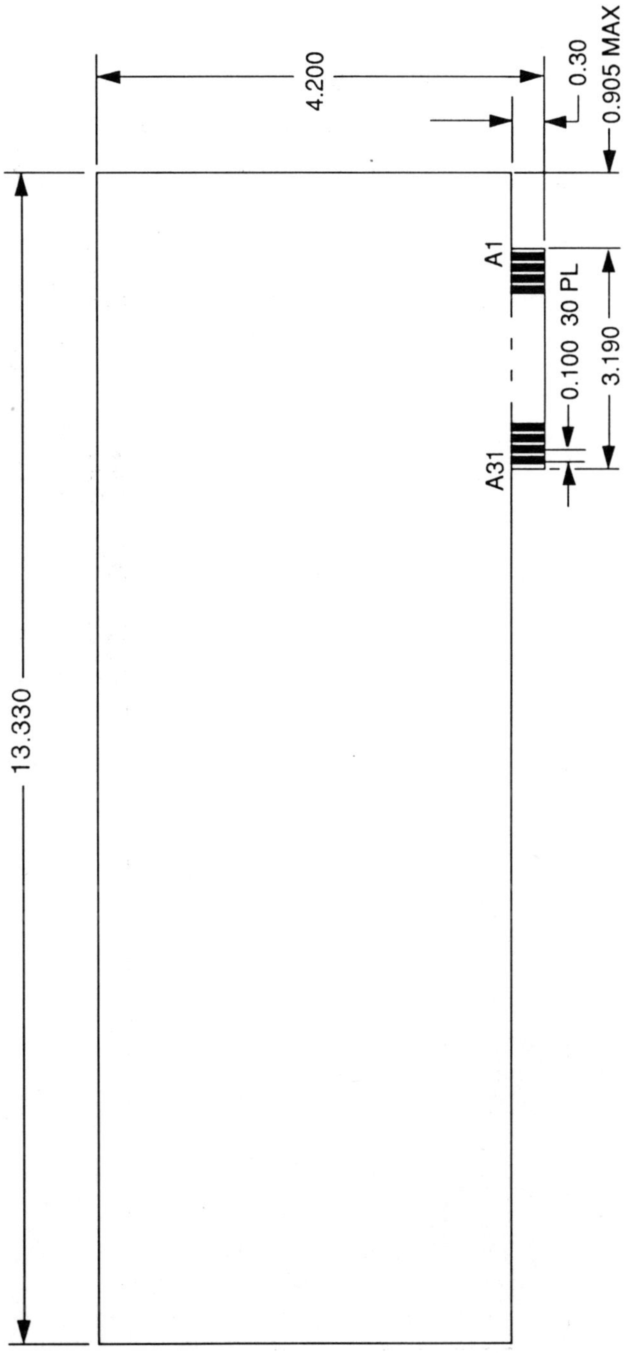

FIGURE 14-2: ISA 8 BIT ADD-ON CARD CONFIGURATION

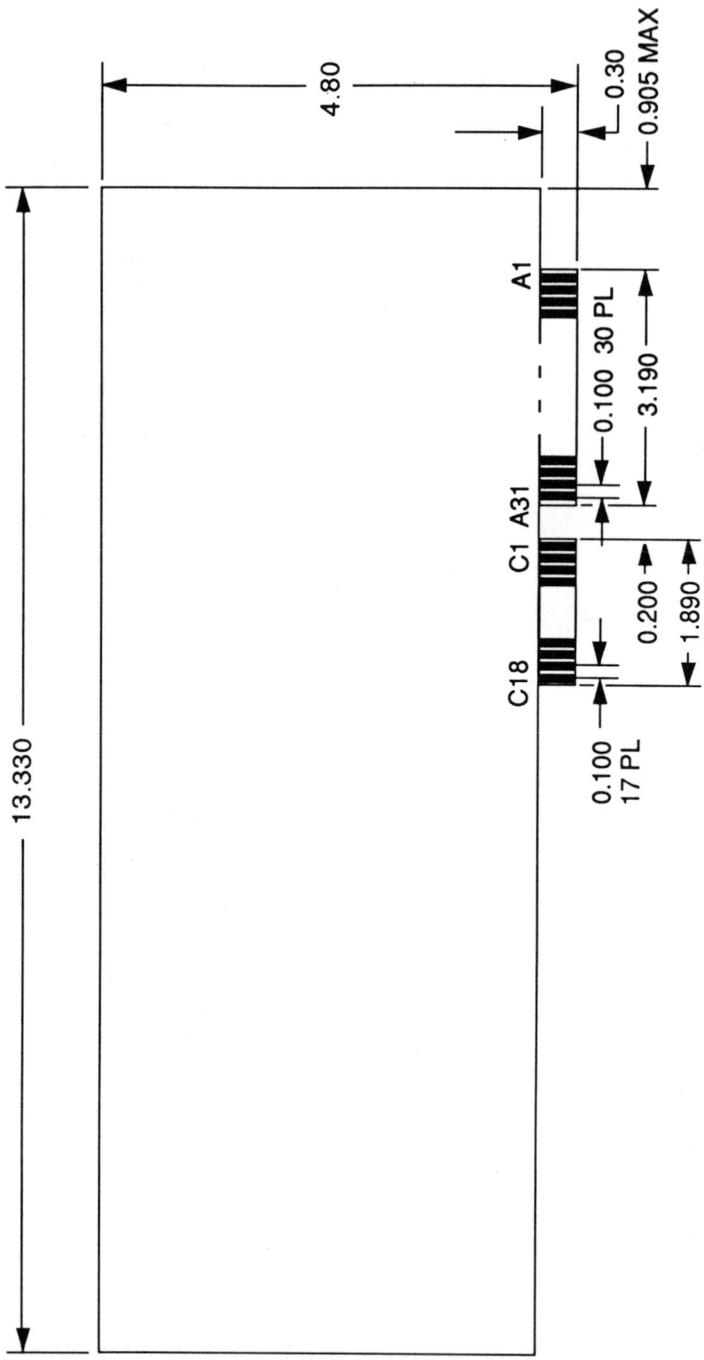

FIGURE 14-3: ISA 8/16 BIT ADD-ON CARD CONFIGURATION

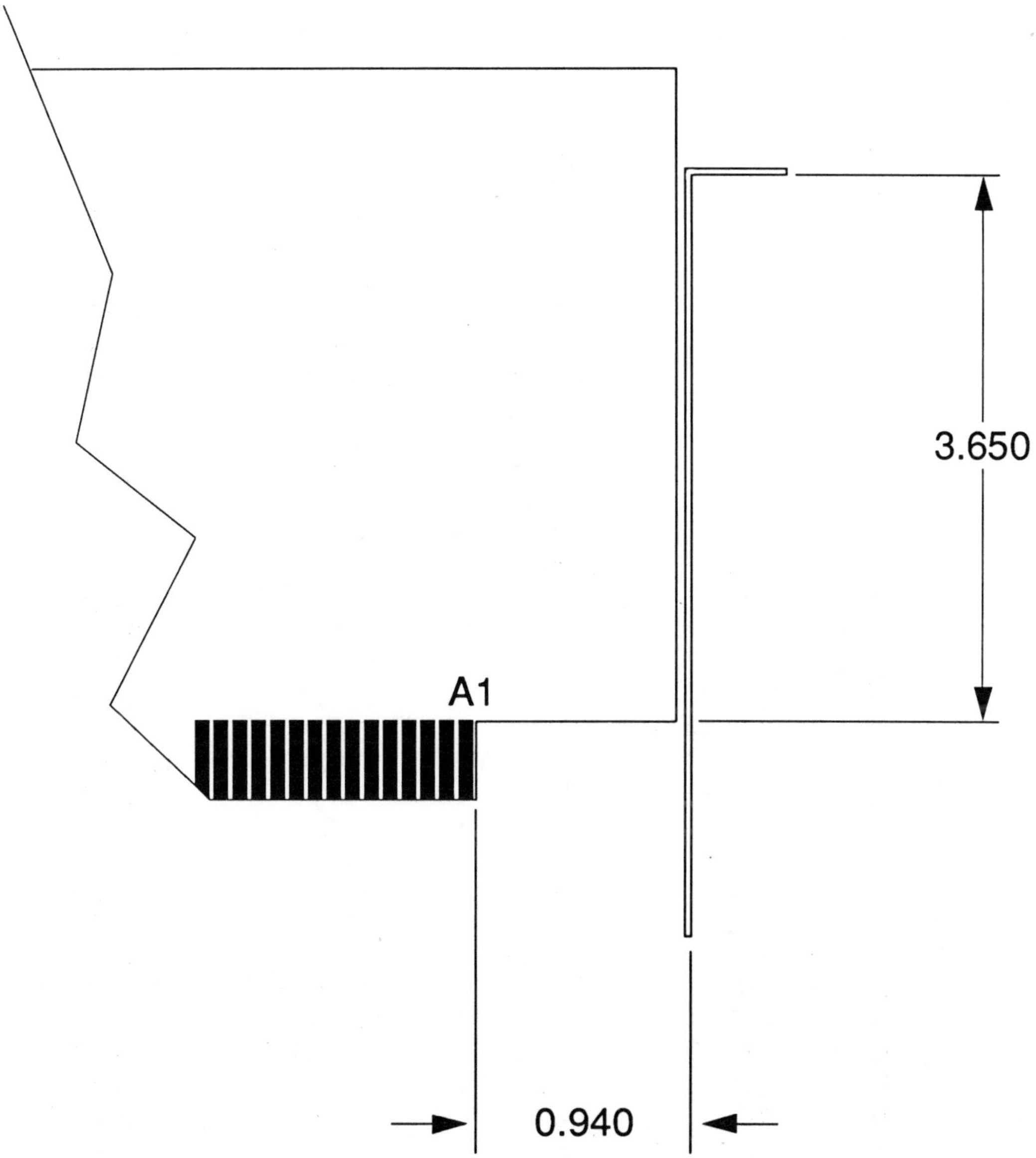

FIGURE 14-4: ISA CARD BRACKET MOUNTING POSITION

Pin Number	Row A	Row B	Row C	Row D (Optional)
1	0V	IOR*	0V	UDP1
2	+5V	IOW*	+5V	UDP2
3	+12V	MEMR*	-12V	UDP3
4	BALE	MEMW*	SBHE*	UDP4
5	SA0	LACHn	SA1	UDP5
6	SA2	0V	SA3	UDP6
7	SA4	SA5	SA6	UDP7
8	SA7	SA8	SA9	UDP8
9	SA10	SA11	SA12	UDP9
10	SA13	SA14	SA15	UDP10
11	SA16	LA17	LA18	UDP11
12	LA19	LA20	LA21	UDP12
13	LA22	LA23	IRQ3	UDP13
14	IRQ4	IRQ5	IRQ6	UDP14
15	IRQ7	IRQ9	IRQ10	UDP15
16	IRQ11	IRQ12	IRQ14	UDP16
17	IRQ15	IOCHCK*	DRQ0	UDP17
18	DRQ1	DRQ2	DRQ3	UDP18
19	DRQ5	DRQ6	DRQ7	UDP19
20	DACK0*	0V	DACK1*	UDP20
21	DACK2*	DACK3*	DACK5*	UDP21
22	DACK6*	DACK7*	TC	UDP22
23	SD0	SD1	SD2	UDP23
24	SD3	+5V	SD4	UDP24
25	SD5	SD6	SD7	UDP25
26	SD8	SD9	SD10	UDP26
27	SD11	SD12	SD13	UDP27
28	SD14	SD15	IOCHRDY	UDP28
29	AEN	MEMCS16*	IOCS16*	UDP29
30	RESET	SRDY*	MASTER*	UDP30
31	+5V	REFRESH*	+5V	UDP31
32	0V	SYSCLK	0V	UDP32

FIGURE 14-5: PC/96 BUS PIN IDENTIFICATION

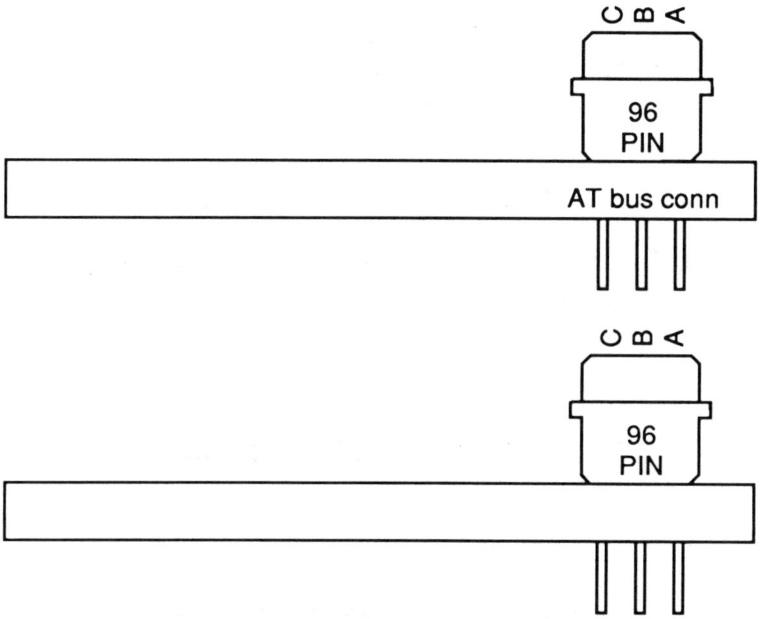

Board Stacking

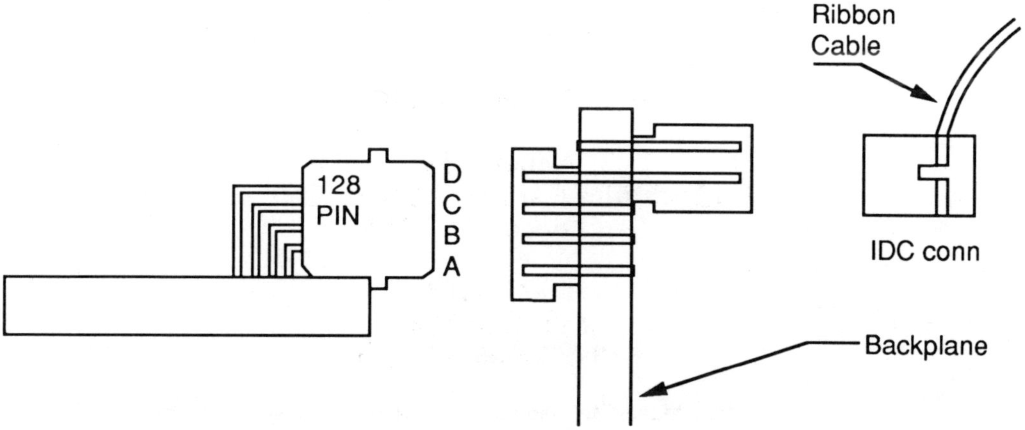

Through-Backplane Connections

FIGURE 14-6: PC/96 BACKPLANE AND STACKING METHODS